INVISIBLE FIXES

INVISIBLE FIXES

How Nanotechnology, Global Science, and Brain Circulation Can Heal Our World

CHRISTIAN ISALOMBOTO NKANGA

Foreword by Twan Lammers

Invisible Fixes

First Edition
2026

Mansfield, Texas, USA

Author: Christian Isalomboto Nkanga

Foreword: Twan Lammers

Cover Design and Typesetting: Louise Brigenshaw

Editing: Joe Pierson (Developmental and Copyediting)
Wayt Gibbs (Editorial Consultant)

Publisher: Independently published

ISBN: 979-8-234-05805-8

Library of Congress Control Number (LCCN): 2026909938

Contact

For permissions, inquiries, or speaking engagements:
christian.nkanga@unikin.ac.cd; stianolnkanga@gmail.com

Disclaimer

For Corentin

In Memoriam

To my late mentor, Prof. Rui M. W. Krause (1970–2024)–
who welcomed me, supported my development, and introduced me to nanotechnology at Rhodes University in South Africa. As a professor of organic medicinal chemistry and nanomaterials, his curiosity, generosity, and commitment to innovation shaped my thinking and contributed meaningfully to my academic journey.
May he rest in eternal peace.

"The real scandal of our time is not that science has no answers–it is that its answers do not reach everyone."

FOREWORD

Let me start with a confession: I don't read books. I love books, though, and I have many of them. I just don't find the time (and/or peace of mind) to read them. I have bought at least 100 books in the past five years, and started almost all of them, but finished none. That said, I did finish 80-90% of *Invisible Fixes*. And I truly enjoyed it, for reasons indicated below.

There are not many books or texts that I (semi-) read that besides describing progress in science, also situate science and its implications in a real-world scenario, particularly in the realities of the world it aims to serve. *Invisible Fixes* is such a book. Dr. Nkanga takes us on three intertwined journeys: a scientific, a personal, and a global one. From his early days at the University of Kinshasa, Democratic Republic of the Congo (D.R. Congo), through formative experiences in Southern Africa, Europe, and the United States, and ultimately back to his home country, D.R. Congo, *Invisible Fixes* not only traces the trajectory of an individual scientist, but it also showcases the importance of always keeping a global perspective.

What stands out already at the beginning of the book is the clarity with which Dr. Nkanga communicates. This clarity is not accidental. It reflects both rigorous scientific training and the influence of the outstanding mentorship he has received during his career. Complex ideas in nanomedicine and drug delivery are distilled into language that is precise, accessible, and purposeful. As someone who has spent much of his career working on nanomedicine, which by many is depicted more like science-fiction and fiction-science than a real medical discipline, I realize how difficult this balance is to achieve. Yet, the clear language and engaging, encouraging storytelling make *Invisible Fixes* appealing and accessible to both nanomedicine specialists and a broader, diverse global audience.

Dr. Nkanga's story is rooted in identity as much as in science. Interestingly, he notes that his surname derives from "Nganga", which means healer in his native Kikongo language, i.e. one who uses plants and knowledge to bring health. This notion resonates throughout the book. It is reflected in his early academic excellence, driven by a belief–and reinforced by mentors–that hard work is the most reliable path forward. It is reflected in his persistence across continents and disciplines. And it is reflected most powerfully in the personal experiences that shaped his scientific development and direction.

Few passages are as striking as those describing the shared burden of hepatic disease within his family. While he himself ultimately gained access to advanced care in the United States–including treatment with targeted drugs similar to the ones he has been working on–his father and brother back home in D.R. Congo did not. This contrast is not presented for dramatic effect; it is presented as evidence, and it leads to one of the most important questions in the book: Are the limits of modern medicine scientific, or are they largely geographic and economic?

This realization reframes discussions on the added value of nanomedicine. Or of any type of advanced medicine. Technologies such as liposomes, micelles, antibodies, nanoparticle vaccines, CRISPR-based gene editing agents, et cetera, all clearly hold enormous promise. They enable the use of difficult drugs (like mRNA), and help to improve pharmacokinetics, biodistribution and therapeutic efficacy, while reducing toxicity. They are being explored not only for cancer, but increasingly also for inflammatory and infectious diseases, incl. leishmaniasis and viral infections. Dr. Nkanga presents these developments with clarity and relevance, demonstrating how they may one day serve to solve global health problems.

Invisible Fixes makes an important point: innovation alone is not enough. Global initiatives can surely catalyze progress, but effective translation and sustained clinical, societal and economic impact require dedicated involvement of the pharmaceutical industry, particularly for processes like large-scale manufacturing, regulatory navigation and distribution logistics. Dr. Nkanga argues, as we (and several others) did in the past, that translation must start with the end-user in mind. Accordingly, he suggests training "science translators", i.e. individuals capable of bridging the complex chain from discovery to delivery, including documentation, manufacturing, regulation, clinical adoption, and financing. This is an excellent suggestion.

In the past 20 years, I have had the privilege of working with many talented and particularly very highly motivated scientists from around the world. This includes students and trainees from underprivileged regions. Supporting highly motivated individuals from such backgrounds is something I deeply value, not only out of principle, but because it strengthens science itself. In this context, Dr. Nkanga's concept of "brain circulation" versus "brain drain" is interesting. Knowledge should not simply move away from where it is generated and needed. It should circulate, leave, return, and ideally eventually contribute to generating strong local capacity. This notion is closely tied to one of the central messages of the book: that science–when efficiently shared, translated, (re)circulated and eventually locally rooted–is the best tool for global repair and long-term societal impact.

Along the same lines, *Invisible Fixes* also showcases a reality that is often overlooked, i.e. that inequity is not at all confined to low-income countries. Even in highly developed countries, such as the United States, access to advanced medicines tends to be limited to those who can afford them. Geography matters, but money matters more. In the end, regardless of how beautiful, elegant and important science is, everything always comes down to money. This is not a new notion. And this will not be easy to resolve.

Books cannot offer simple solutions for complex problems. However, they certainly help to sharpen the questions, clarify the challenges and inspire actions. They remind us that the future of (nano-) medicine will not be determined by what we discover, and what we publish, but by whether we learn to share, translate and deploy those discoveries in a timely, equitable and financially viable manner.

Reading *Invisible Fixes* was inspiring to me. And it will be for you. It provides personal and general insights that help to contextualize what we do in (nano-) medicine research, and why we do it, and what we should strive for. It takes us from moving moments in a tiny municipality in Mbandaka, D.R. Congo, to the smallest dimensions of state-of-the-art medical materials research, and to global pathways for scientific training, knowledge-exchange and medical impact. Ideally, we would live in a world where health, healers ("Ngangas"), and healing do not depend on where one is born, or what one can afford. I sincerely hope we will all, together, be able to visibly fix that!

Professor Dr. Dr. Twan Lammers

Department of Nanomedicine and Theranostics, RWTH Aachen University, Germany

AUTHOR'S NOTE

"Christian, the doctor says it's too late. There is nothing he can do for me. I have cancer."

My older brother, Jean Corentin Nkanga, said these words one evening in December 2019 in Kinshasa, Democratic Republic of the Congo (DRC). We had just returned from a clinic with no treatment plan, no options, no relief–only the sound of his voice breaking under pain.

Days later, unable to endure the suffering, he pleaded with me, "Let's go back. There must be something he can do–anything." At the hospital, the only intervention offered was intravenous fluids. My brother clung to the drip as if it were a lifeline, believing it might shrink the mass that filled his abdomen. When the infusion ended, the doctor returned only to repeat what we already knew: The disease was too advanced. Nothing more could be done.

We went home defeated. I watched cancer consume a man who had sacrificed everything for me after our father's death in 2006–a brother who carried me through my undergraduate studies, who believed in my future when resources were scarce. On the morning of January 23, 2020, I received the call that he had passed away.

What haunted me was not only the loss, but the silence of science.

How could modern medicine have no answers?

Would my brother have lived if he had been treated elsewhere?

Were the limits of medicine scientific–or were they geographic, economic, and structural?

Only weeks later, in February 2020, I boarded a plane to California to begin a postdoctoral fellowship at the University of California, San Diego. The contrast was impossible to ignore. After my doctoral training at Rhodes University in South Africa, I entered an environment where advanced biomedical nanotechnologies were not theoretical–they were active, funded, tested, and translated. I was engineering nanoparticles against cancer and emerging infectious diseases, witnessing firsthand solutions that existed just beyond the reach of those who needed them most. Some of the discoveries I participated in during those years–liposomes, plant viruses, hydrogels, green metallic nanoparticles–later became the scientific backbone of this book, rewritten here not as journal articles, but as human stories of how medicines are engineered.

Then life turned again. In San Diego, I was diagnosed with a serious liver disease. Unlike my brother, I had options. Standard treatments existed, though not without risks. But I was eventually offered a smart, targeted nanomedicine–one that used less than 10 percent of the conventional drug dose, eliminated long-term toxicity, and delivered precise therapeutic action. I do not share this as medical advice, but as lived evidence of what becomes possible when advanced therapies and expert systems are within reach. I survived because invisible fixes were available.

That experience gave a name to what I had been living.

By *invisible fixes*, I mean three interconnected forces. First, nanotechnology–tools engineered at a scale too small to see, yet powerful enough to transform medicine, diagnostics, and environmental health. Second, brain circulation–the movement of knowledge, skills, and people across borders that allows science to travel, adapt, and serve societies beyond where it is first developed. And third, translational capacity–the systems, expertise, and institutions required to turn discovery into real-world tools rather than leaving science trapped in publications. Together, they form a quiet revolution: solutions that are unseen, yet capable of changing who gets to live.

This book is written for scientists and nonscientists alike. I explain nanotechnologies using accessible language, stories, and metaphors–so that readers without a technical background can understand how nanomedicines work, why they are different from conventional drugs, and why they matter. Along the way, I take you inside four real examples from my own research journey: teaching tuberculosis drugs where to go, reengineering plant viruses into cancer medicines, building hydrogels that make immunity last longer, and turning plant waste into antimicrobial nanomaterials in the age of drug resistance. But this is not only a book about technology. It is a book about equity. It is also a book about responsibility. Every powerful medical technology carries risk, and one of the most neglected questions in public conversations about nanomedicine is safety: what the body tolerates, what it rejects, and what long-term exposure may cost. That is why this book includes a chapter on nanotoxicology–because hope without safety is not progress.

The loss of my brother in Kinshasa and my survival in San Diego transformed my life's mission. Science cannot remain the privilege of geography. If nanomedicine could save me, then it must be made accessible and affordable for others–regardless of where they are born.

In the years since, I have worked to translate that conviction into action. As a professor at the University of Kinshasa, I helped build nanotechnology laboratories from the ground up, leading to the creation of the *Centre de Recherche en Nanotechnologies Appliquées aux Produits Naturels (CReNAPN)*–the first nanotechnology research center in the DRC. As a pharmaceutical development scientist and later chief scientific officer in industry, I learned firsthand how hard it is to move from discovery to a regulated product–and why so many promising innovations never reach patients. And through leadership roles in the DRC Young Academy of Sciences and the Global Young Academy, I have advocated for mobility, collaboration, and fairness in global science.

Rather than leaving permanently, I chose to circulate–moving knowledge back and forth between Kinshasa, South Africa, Europe, and the United States. I have seen how collaboration dismantles barriers, how science grows stronger when shared, and how capacity can be built where it is most needed.

This book tells that story. From Mbandaka to the nanoscale. From personal tragedy to collective responsibility. From silence to solutions.

We live in an era of shared vulnerability: pandemics, climate disruption, antimicrobial resistance. These threats do not respect borders, and neither can the solutions. To confront them, science must be both innovative and equitable. Nanotechnology, global science, and brain circulation are not luxuries–they are necessities. Because the future of health will not be determined only by what we can invent, but by what we can translate, scale, and share.

This book is my invitation to you: to see how the smallest things can carry the greatest hope, and how invisible fixes can help heal our world.

CONTENTS

INTRODUCTION **23**

PART I - THE JOURNEY: FROM MBANDAKA TO GLOBAL SCIENCE **29**

Chapter 1 Roots in Mbandaka **31**

Chapter 2 Crossing Borders **40**

Chapter 3 Becoming a Scientist of the World **50**

PART II - THE SCIENCE: NANOTECHNOLOGY & HEALING **61**

Chapter 4 Invisible Giants: The Power of Nanoparticles **63**

Chapter 5 Nanomedicines in Action: When the Invisible Becomes Clinical **73**

Chapter 6 Engineering Time: Nanotechnology and the Rise of Long-Acting Medicines **97**

Chapter 7 Green Nanotechnology: Healing with What the World Throws Away **110**

PART III - FROM DISCOVERY TO FIXES **311**

Chapter 8 Green Silver, Living Chemistry: When Plants Don't Just "Make Nanoparticles," They Program Them **121**

Chapter 9 Teaching Medicines When to Act: Nanotechnology, Tuberculosis, and the Ethics of Arrival **134**

Chapter 10 When a Virus Becomes a Medicine: Reengineering Nature to Fight Cancer **145**

Chapter 11 When Time Becomes the Medicine: Hydrogels, Viral Nanoparticles, and the Architecture of Sustained Immunity **153**

Chapter 12 Beyond the Bench: Building Translational Capacity Where It Matters **163**

Chapter 13 Nanotoxicology: Making Sure Invisible Fixes Do Not Become Invisible Harms **173**

PART IV - THE VISION: BRAIN CIRCULATION AND GLOBAL HEALTH **311**

Chapter 14 The Brain Drain Dilemma **189**

Chapter 15 Brain Circulation and Global Science as an Invisible Fix **199**

Chapter 16 Building Equitable Science **206**

EPILOGUE – MAKING THE INVISIBLE INEVITABLE **217**

ACKNOWLEDGEMENTS **221**

GLOSSARY: UNDERSTANDING THE LANGUAGE OF INVISIBLE FIXES **224**

KEY READINGS **244**

INTRODUCTION

The Power of What We Do Not See

Most of the forces that shape human life are invisible.

Electricity moves silently through walls. Data crosses oceans in fractions of a second. Bacteria rewrite the rules of medicine without announcing themselves. And today, particles thousands of times smaller than a grain of sand are redefining how diseases are treated, how materials are built, and how life itself can be protected or repaired. Some of these technologies do not just deliver medicine–they control time itself: how long a drug lasts, when it activates, and how the immune system remembers.

We live in an era where the smallest things increasingly determine the largest outcomes.

Nanotechnology–science at the scale of atoms and molecules–belongs to this quiet category of power. It does not announce itself in daily headlines. It cannot be seen by the naked eye. Yet it already sits inside cancer therapies, vaccines, tuberculosis drug delivery systems, wound dressings, diagnostic tools, water purification systems, and antimicrobial materials–sometimes as the active solution itself, and sometimes as the hidden engineering that makes a therapy safe enough to exist. It influences whether a drug heals or harms, whether a treatment reaches the right tissue or the wrong one, whether innovation becomes affordable or remains locked behind borders and balance sheets.

At the same time, another invisible force is shaping our world just as profoundly: the movement of human knowledge.

Ideas migrate. Skills travel. Scientists cross continents. Expertise accumulates in some places while disappearing from others. This flow of talent–sometimes called brain drain, sometimes diaspora, sometimes globalization–quietly determines which societies solve problems and which continue to endure them.

These two invisible forces–nanotechnology and the circulation of human knowledge–rarely appear in the same conversation. One is treated as technical; the other as political or economic. Yet together they form a single question that defines our century: Who benefits from scientific progress–and who is left waiting?

This book is about that question.

In the center of the book, the question becomes practical. I take readers inside several of my own peer-reviewed discoveries–not as journal articles, but as human stories of how science is engineered: liposomes designed to teach tuberculosis medicines where to go, plant viruses reprogrammed into cancer therapies, hydrogels that turn time into immunity, green silver nanoparticles that challenge antimicrobial resistance, and the safety science of nanotoxicology that determines whether these innovations can ever be trusted.

It is about how scientific breakthroughs are born, how they travel, how they stall, and how they can be redirected. It is about why extraordinary medical technologies exist alongside preventable suffering. Why advanced therapies flourish in some hospitals while entire regions rely on outdated tools. Why innovation accelerates globally, yet access remains profoundly unequal.

But this is not a book about despair. It is a book about possibility.

It argues that the future of medicine, public health, and environmental protection will not be shaped by technology alone–but by how knowledge moves, how institutions are built, and how deliberately societies choose to share their scientific power. And it argues that sharing science is not only about publishing papers–it is about building translational capacity: the people, systems, and institutions that turn discoveries into usable medicines, diagnostics, and public-health tools.

Nanotechnology offers new tools: smarter drugs, targeted therapies, sustainable materials, and medical platforms capable of doing more with less. Brain circulation offers a complementary solution: scientists who do not vanish from their home countries but connect worlds–translating global advances into local capacity, and local challenges into global research priorities.

Together, they form what this book calls invisible fixes: nanoscale tools and global knowledge systems that operate beneath the surface, yet determine who gets to benefit from modern science.

The chapters that follow will move between science and society, laboratories and classrooms, hospitals and policy rooms, developed and developing worlds. Because in the end, science does not fail only in laboratories–it fails when the world cannot carry it across borders, into clinics, and into the lives of those who need it most. Chapters will explore:

- How clinically approved nanomedicines already reshape cancer therapy and infectious disease treatment

- How liposomes can be engineered to deliver tuberculosis drugs more intelligently

- How viruses can be redesigned into cancer immunotherapies and light-activated treatments
- How hydrogels can turn time into medicine by sustaining immunity and reducing dosing burden
- How green nanotechnology can turn waste into lifesaving materials in the age of AMR
- Why long-acting injectable drugs depend on nanoscale engineering
- Why nanotoxicology–safety at the nanoscale–determines what can reach real patients
- Why scientific migration can weaken nations or strengthen them, depending on how it is structured
- And how equitable science can be designed, funded, translated, and sustained

This is not a technical manual. It is not a manifesto. And it is not a memoir alone. It is an attempt to map the hidden architecture of modern progress–and to show how it can be rebuilt so that scientific advancement no longer belongs to a few places, a few institutions, or a few lives, but becomes a shared instrument for human dignity.

If the twentieth century taught us how to split the atom, the twenty-first century must learn how to distribute its benefits.

That is the journey this book invites you to take: into the smallest dimensions of matter, into the global pathways of knowledge, and into a future where healing does not depend on geography.

Many of the solutions are already here–or already possible.

They are simply too small–and too quietly moving–to be seen.

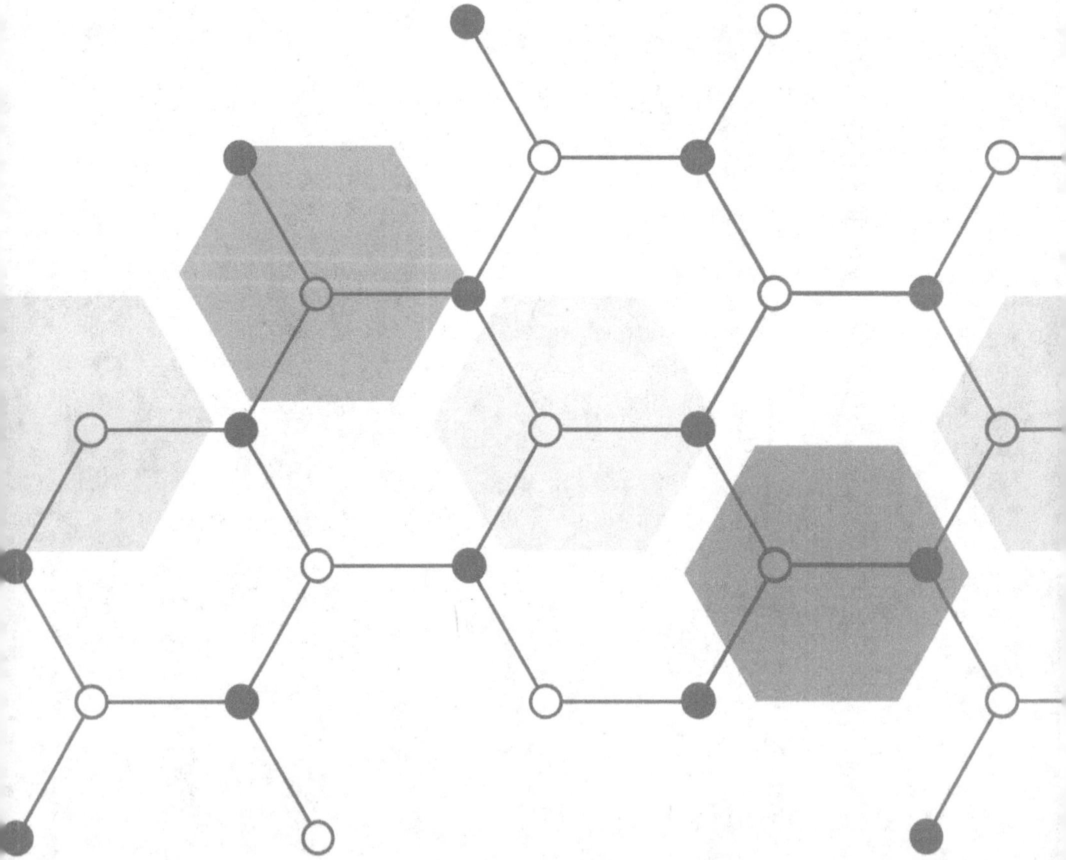

PART I

THE JOURNEY:

From Mbandaka to Global Science

CHAPTER 1

ROOTS IN MBANDAKA

Introduction

The Congo River and the Congo Basin Rainforest were both our pride and our peril. The river's waters fed us with its delicious fish and carried traders and stories from far-off places. The rainforest, vast and rich, was celebrated for its astonishing biodiversity, its medicinal plants, and its agricultural bounty. Yet both also carried disease.

As a child in Mbandaka–and during visits to my mother's town, Bikoro–I saw epidemics sweep through communities. Cholera. Monkeypox (today called mpox). Malaria. These diseases claimed lives with terrifying speed. I learned early that survival was fragile, that life in our region could be as precarious as it was abundant.

My grandfather was a nurse, and his work inspired me. I grew up fascinated by the healing power of plants and the resilience of traditional medicine. I was told that my surname, Nkanga, came from the word *Nganga* in my native language–meaning healer, one who uses plants and knowledge to bring health. That heritage shaped how I saw myself: part of a lineage where science and healing were already intertwined, long before I ever set foot in a laboratory.

But life was full of sudden lessons. One day, my grandmother, who had seemed perfectly well, collapsed while doing her household chores. Within hours, she slipped into a coma, and by nightfall, she was gone. No warning, no treatment, no explanation.

Her death haunted me. How could someone so alive vanish so suddenly? Were those who looked healthy truly safe? Could anyone in my family disappear the same way? Was this our fate in Congo, or did people elsewhere in the world live under different rules?

Those questions stayed with me. They became the first spark of my journey–a relentless curiosity to understand the mysteries of the human body, the hidden causes of disease, and above all, the injustice of survival being determined by where one is born.

Congo River and Rainforest: Lifeline and Paradox

Born on May 16, 1989, in a small municipality of Mbandaka, I grew up surrounded by two majestic forces of nature: the Congo River and the Congo Basin Rainforest. Like many Congolese, I learned to see these two as both gifts and challenges–lifelines that sustained us and forces that tested our resilience.

Peter Forbath, in his 1977 classic *The River Congo*, described the river as "not merely a river; it is a continent's beating artery, commanding awe for its immensity and commanding respect for the life it sustains." Having spent the first seventeen years of my life close to this river, I can testify to its central role in shaping the rhythms of our daily existence.

During the dry season (June through September), the river receded and sand islands emerged, transforming into vibrant fishing camps. My brothers, friends, and I would walk along the riverbank, collecting fish by hand–no nets needed. Fish was so abundant that the dry season felt almost festive. Yet, the same period often meant dry boreholes, forcing families to fetch water from distant springs. I used to volunteer to bring water not only for my family but also for some of the neighbors in need of help. With my wheelbarrow, I would walk nearly five miles to a spring, fill several five- to twenty-five-liter containers, and return with them balanced carefully. One day, my stepmother, Wyvine Bofenda, met me halfway and found me hauling seventy-five liters of water. "Do you want to break your back?" she exclaimed. On the way back, I delivered extra bottles to their owners, proud of being useful to my community.

The rainy season, by contrast, brought abundant water but made fishing difficult. Fish camps disappeared underwater, and fishing relied on traps and nets with unpredictable results. With my maternal grandfather, Henry Boongongo, we set eel-bucks (a traditional boxlike fish trap–often woven from wood, reeds, or wicker–used by river communities) in rivers to catch eels, but success was rare. Instead, we depended on professional fishermen, who would push carts loaded with fresh fish–especially juvenile alevins sold cheaply in bulk, a beloved staple for local families. Yet, the rainy season also meant flooding, infectious diseases, and mosquitoes, reminding us how fragile abundance could be. Floodwater was never considered safe for drinking, even though it often turned the neighborhood into an open playground where we spent hours playing soccer with friends. Rainwater, by contrast, was used for daily life–but only after caution and patience. We learned early that it had to be left undisturbed for some time so that suspended particles could settle before the water was used for drinking or household activities. Once the water had settled, the level of contamination became unmistakable: a dark layer of particles collected at the bottom of the container. As children, we believed these impurities came mainly from dust accumulated on the roof, since rainwater was collected by placing buckets under the flow running off the roofing sheets. To reduce this contamination, we followed an unspoken rule: never collect rainwater during the first few minutes of rainfall. We waited deliberately, allowing the initial wash of dust and debris to run off the roof, and only then positioned multiple buckets beneath the water stream as it poured down.

My experiences with the rainforest came mostly during visits to my mother's home village, Bongonde (Bikoro Territory), where I stayed with my grandmother, Angelika Bisaka. At age six, during a school break, I discovered the rhythms of subsistence agriculture. My grandmother cultivated cassava fields deep in the forest. Each day at 5 a.m., she would leave with a basket and return hours later, carrying cassava leaves and tubers. She soaked tubers in spring water pools to soften them, then cooked some into cassava bread and dried others to make flour for *fufu*. The process was both routine and ingenious, using every part of the plant and relying entirely on natural resources.

Fufu was one of the most familiar foods of my childhood–simple, heavy, and comforting in a way that felt almost universal in our home. It was made by gradually mixing a fine flour from milled tubers–most often cassava, sometimes yam or plantain–into hot water, then stirring and folding it with strength and patience until it became smooth and elastic. When it was ready, it sat like a warm, pale mound on the plate, soft enough to shape by hand. We ate it without forks, tearing off small pieces and using them to scoop up soup or sauce. It was not a luxury meal, but it was reliable–and in a life where many things were uncertain, fufu was one of the quiet constants that held us together.

The spring water in the village was crystal clear and cool. We drank it from traditional calabash cups, which kept it fresh. After so many years, I still remember the amazing moments in Bongonde. Monkeys darted through trees, frogs croaked at dawn, and the forest sounded alive, almost breathing. For us, the river and rainforest were providers–of food, water, trade, and identity–but they were also sources of disease and hardship.

Early Lessons in Resilience

Growing up in Mbandaka meant constantly navigating between abundance and scarcity. Markets near river ports bustled with traders arriving by boat. At Basoko Market, I saw travelers bringing bread and fish from surrounding villages every Thursday evening. As teenagers at Nsong'a Liyanga College, my friends Jules Bakombo, Henry-Thomas Dilangu, and I used to explore downtown ports to admire clothes shipped from Kinshasa. We saved to buy stylish outfits, wanting to look like "big-city boys."

Yet, amid this vibrancy, disease outbreaks were common. Equateur Province was regularly hit by cholera from contaminated water, mpox (monkeypox) outbreaks, and mosquito-borne illnesses like malaria and chikungunya. Floodwaters often left puddles that became breeding grounds for mosquitoes. Modern medical facilities existed, but many communities relied primarily on traditional healers and medicinal plants.

I experienced this firsthand. After fracturing my left arm during an acrobatic game, my family brought me to a local healer. She warmed medicinal leaves, added palm oil, wrapped my arm, and replaced the bandages every other day. I recovered fully without ever seeing a hospital. Later, when I suffered from swollen tonsils, I followed a traditional treatment and incantation ritual: applying a slurry made from burnt banana peel ashes mixed with water, then singing in front of a widow's mirror in my village language: "*atama tama lodulwa lonzemba.*" (Meaning: "I am here in the house of a widow.") After several sessions, the swelling disappeared.

Christian I. Nkanga in Mbandaka, Equateur Province, D.R. Congo (2001). Photo credit: Pierette E. Nkanga

Chickenpox was treated with salt water, and conjunctivitis was cured by collecting early-morning dew from plant leaves and squeezing droplets into the affected eye. Skin conditions like pityriasis were treated by rubbing fresh leaves of *Cassia alata* (the "candle bush") on the skin. These practices were effective, accessible, and deeply embedded in our community life. They didn't replace modern medicine; they preceded it.

These experiences seeded in me a deep respect for local knowledge and a curiosity about how healing actually worked. How could plants and simple rituals heal infections or injuries? Were these effects purely empirical, or was there a biological explanation hidden beneath tradition?

My paternal grandfather, Joseph Isalomboto Nkanga, was a nurse practitioner. He kept medical tools at home and treated wounds for neighbors. People often told me I not only carried his name but also resembled him. Watching him clean, debride, and dress wounds ignited my early desire to become a medical doctor.

When I told him of my ambition, he smiled and said, "This is a good choice; it reflects the meaning of your surname." Curious, I asked what "Nkanga" meant. He explained that in our village tongue, "Nkanga" derives from "Nganga," meaning *someone who heals using plants.* Learning this gave my childhood fascination with healing a deeper cultural and personal dimension. It felt like destiny and responsibility intertwined.

A Sudden Loss and Questions about Life

One day, while visiting my maternal grandmother in the neighboring municipality, Mbandaka II, she collapsed unexpectedly and fell into a coma. She died later that day; yet, until then, she had appeared healthy. I was shocked. How could someone go from working in the morning to passing away at night without warning?

This tragedy stirred existential questions in my young mind. Are those who look healthy truly healthy? Why do some lives end suddenly while others endure? Does the place where you are born determine your chances of survival?

As a fourteen-year-old boy, I didn't have answers to these questions, but I began to see health and survival as fragile–not guaranteed, but conditional on factors I didn't yet understand. These unanswered questions later became a driving force behind my scientific curiosity.

Education, Determination, and Invisible Legacies

After two years at Nsong'a Liyanga College, I transferred to Frère Iloo Institute to focus on science. My dream of becoming a medical doctor drove me to study intensely. My hard work paid off. By my second year, I was top in mathematics and among the five best students in my class.

Yet, tragedy struck again. My father, Steward Isalomboto Nkanga, suffered from chronic liver disease for years. In 2006, at age sixteen, I watched my family mobilize pastors and priests to pray for him, believing the illness was spiritual. Eventually, he traveled to Kinshasa for medical care. On September 13, I heard cries from outside and saw my elder brother weeping. My father had died. He was everything to me–the pillar who encouraged us to study as the ultimate inheritance. In Lingala, he often said, "*Petits, botanga classe. Yango nde eleki tout. Nionso wana eza pamba.*" (Meaning: "Boys, stick to your studies; it is above all. Everything else is nothing.")

We didn't inherit buildings or wealth, but we inherited a legacy of education–an invisible gift that would become the foundation of my journey beyond Mbandaka.

Looking Back: Seeds of Invisible Fixes

When I think of my childhood now, I see more than memories–I see patterns. The river and rainforest shaped our lives. Traditional healers showed me that unseen mechanisms could heal. Family legacies and sudden tragedies taught me resilience and the urgency of seeking answers.

At that time, I had never heard the word *nanotechnology*. But I already believed that some solutions are invisible to the naked eye, hidden in places we might not think to look. Chapter 1 is where the questions began–questions that would later lead me from Mbandaka to Kinshasa, from Kinshasa to South Africa, and eventually to the laboratories overseas.

CHAPTER 2
CROSSING BORDERS

Introduction

Grief is a river. It can drown you–or carry you to new shores. In 2006, when my father passed away, the current of loss swept through my world, threatening to sink my childhood dream of becoming a medical doctor. Yet that same current carried me beyond the familiar banks of Mbandaka, down the Congo River to Kinshasa, through demanding years of study, into the laboratories of South Africa, and eventually across the Atlantic to San Diego. What began as tragedy became the journey that shaped my search for answers science had not yet given my family.

Redefining the Dream

When my father died on September 13, 2006, the dream I had carried for years–to become a medical doctor–began to fade. I was completing my final year of high school, preparing to enter university the following year. But my father had been my only financial supporter. Ten years of medical studies now felt impossible. I started to look for a shorter, more affordable path.

One conversation shifted everything. Serge Nzinga, a friend of my brother and a respected elder from Nsong'a Liandja College, said to me, "Christian, you can also study pharmacology."

At first, I hesitated. Many around me viewed pharmacy as simply selling medicines behind a counter. But when I learned that pharmacists could also transform plants used in traditional medicine into modern drugs–tablets, capsules, syrups–something clicked. I remembered my surname, *Nkanga*, derived from *Nganga*, meaning "healer with plants" in my language. My heritage and curiosity converged. Pharmacy felt like a bridge between tradition and modern science.

A Journey to Kinshasa

In July 2007, after passing my state exams at Frère Iloo Institute, I boarded a boat with my elder brother Jean Corentin Nkanga for a cross-province trip from Mbandaka to Kinshasa, the capital. For someone raised in a small river city, starting undergraduate studies in pharmacy at the University of Kinshasa (UNIKIN) was both intimidating and exciting.

I promised myself–and my family–to work relentlessly. My late brother, Jean Corentin, and cousins like Natacha Iyongo supported me with pride. Others were skeptical about the toughness of the program.

Early on, a friend of my cousin Darly N. Iyongo, named Anicet Kanana, introduced me to his cousin, a former pharmacy student who had transferred to the faculty of medicine. His advice was simple and unforgettable: "Only hard work counts. Nothing else."

Motivated, I threw myself into my studies. By the end of my first year, I was one of only eight students to pass the first session out of nearly four hundred. In my second year, I pushed even harder. Alongside my classmate Sephora Mianda, I became one of the first two students in a decade to earn a distinction in pharmacy. Students from across the university came to see us, to shake hands and celebrate. It was a moment that convinced me more than ever that hard work truly pays off.

Hardship and Innovation

The third year brought a brutal financial crisis. My father's small savings had run out, and my brother and I were on our own. I couldn't pay tuition. I wrote dozens of emails to sponsors and organizations abroad, pleading for help; no responses came.

We took hourly jobs in factories in the municipality of Limete (Kinshasa), packaging and labeling bottles. The work was exhausting and paid just enough for groceries, not tuition. Around this time, a scholarship call for the Bourse d'Excellence Bringmann aux Universites Congolaises (BEBUC) opened. I applied but wasn't selected. It felt like the end.

But encouragement came from unexpected places. Dieudonné Tshitenge, one of the BEBUC scholars at that time, told me, "Everything will be all right. You have distinguished; you know what that means at the faculty." Dieudonne knew that distinguished students were automatically offered the position of junior lecturer to support practical work at the faculty. And Professor Karine Ndjoko, one the BEBUC panel members, advised, "Don't give up. Keep coming to campus. Keep working hard."

I was inspired, but reality struck again: No tuition meant no studies. My brother approached Professor Dibungi T. Kalenda, who knew me well. Professor Kalenda paid my outstanding fees and encouraged me to continue. His act of generosity changed the course of my studies.

Later, financial struggles inspired innovation. Remembering the healing properties of Cassia alata leaves used in my childhood for skin diseases, I formulated an ointment for pityriasis, eczema, and psoriasis. I named it ISALOT–after part of my middle name *Isalomboto* and "*loto*" (pityriasis in lingala, my local language).

With Professor Pierre Senga's support (lab space, financing, and contacts), ISALOT production grew. It became my small business, sustaining me through the later years of study. Further support from many family friends also kept me afloat.

During my final year, financial aid waned again. My friend, Sephora Mianda, paid my last tuition fees, allowing me to finish my internship. A temporary job with local firms in pharmaceutical marketing kept me going through graduation.

Discovering Nanotechnology

In June 2014, I obtained my Bachelor of Pharmacy degree with distinction and took the pharmacist's oath. But my mind was already looking beyond. I applied to numerous scholarships abroad (such as Belgium, France, Canada, Australia, Germany, South Africa, Malaysia, etc.) with the support of Professors Dibungi T. Kalenda, Patrick B. Memvanga, and Nadège K. Ngombe, but without success.

One day, browsing Rhodes University's website, I found Professor Rui W. M. Krause's profile. Impressed, I emailed him to request a master's position in his lab. He responded kindly, "I would be delighted to offer you a place but I currently do not have any fellowships for non-South Africa students available. ... attached is the official application form once you have managed to sort something out." He also sent me the contact of the bursary office to enquire about suitable fellowships to apply to.

When BEBUC announced a new call in 2015, I applied again, this time for a master's. I discussed my research idea–plants for sickle cell treatment–with Professor Krause. He suggested a twist. "Let's use nanoparticles, specifically liposomes, to deliver genes to tackle sickle cell disease."

It was the first time I had heard the word *nanoparticles*. I was intrigued. I accepted the proposal. After my oral presentation in July 2015, I was awarded the BEBUC Master's Scholarship. I still remember Professor Gerhard Bringmann's words that night: "Christian, go to South Africa. Work hard and be the best you can possibly be."

South Africa: Growth and Challenge

I arrived in South Africa in December 2015, welcomed at OR Tambo Airport by my cousin Wesly Iyongo. At Rhodes University, I went straight to Professor Krause's office with my bags–I had no plan, no place to stay. Despite his tight schedule, being the head of department, he spent the afternoon driving me around, helping me settle in.

He set me up in a department house and introduced me to Dr. Xavier Siwe Noundou, a postdoc who became a mentor. Professor Krause not only trained me scientifically but also helped me integrate culturally. He took the research group to restaurants, encouraging me to speak English despite my nerves. Once, at the restaurant named Rats and Parrots, he joked, "Christian, it's your turn to speak. If you don't say something, you'll pay the bill today!"

As a native French speaker unfamiliar with English and Xhosa, the culture shock could have been overwhelming. His support made it manageable–even inspiring.

Financial hardship followed me. Back home, my brother's health was deteriorating. Many of my pharmacy school friends often collected money to buy him medicine. When things became desperate, Professor Krause reached into his own savings to help.

Through perseverance, I completed my master's thesis in one year, winning two new scholarships: the DST/NRF Innovation and Priority Research Areas Scholarship (South Africa National Research Foundation, 2017), and the Henderson Doctoral Scholarship (Rhodes University, 2018).

In April 2019, I completed my PhD on nanotechnology for drug delivery and returned to Kinshasa to lecture at University of Kinshasa (UNIKIN), where I had already been appointed junior lecturer in 2018.

Return to Congo and a Brother's Battle

Returning home was harder than expected. I taught at the university for six months without salary, while my brother Jean Corentin entered his final, painful battle with liver disease.

In September 2019, he underwent surgery. The team almost stopped mid-procedure because funds were insufficient. It was devastating. In November 2019, I received an offer to join Professor Nicole Steinmetz's lab at the University of California, San Diego (UCSD) for postdoctoral research in nanoengineering.

The way I bumped into this postdoctoral opportunity at UCSD was memorable. One day in September 2019, I opened my phone and saw that my name had been tagged on Twitter. I clicked out of curiosity. The tag came from Professor Karine Ndjoko, one of my

BEBUC mentors, and she had written under a postdoctoral call shared by Professor Nicole Steinmetz at the University of California, San Diego. Her message was short but powerful. She tagged me and wrote "@Christian, Nature author," pointing to my work and signaling that I was someone worth noticing.

I had never met Professor Steinmetz. I had never spoken to her. But in that moment, I understood the hidden power of networks–how scientific lives can sometimes pivot on a single act of recognition.

I immediately searched for Professor Steinmetz's email address, and I sent her my CV along with a brief message expressing my interest in a postdoctoral position. To my surprise, she replied quickly–and favorably. She suggested that we schedule a meeting.

When the day came, she asked me to present my research, and in particular, to speak about the work I considered most meaningful. I chose the project that had shaped my doctoral identity: the strategy I developed to control the release of an antituberculosis drug from nanoparticles. I explained the problem we were trying to solve–how tuberculosis treatment is long, exhausting, and often toxic–and how controlled release could reduce burden while improving precision. I walked her through the logic, the formulation approach, and the scientific thinking behind it.

She listened carefully, then gave me feedback that I still remember: She told me my experience in nanotechnology for drug formulation was compelling. But she also made something clear–something that would later become one of the central lessons of my own career. Scientific skills alone were not enough. She wanted to see how I wrote, how I structured ideas, and whether I could communicate science with clarity.

She asked me to draft a proposal.

I did it immediately, working intensely, shaping the narrative, refining the argument, and making sure the document reflected not only what I knew, but how I thought. When I sent it to her, her response came quickly again–positive, encouraging, and decisive. She then introduced me to members of her research group. We held a few online meetings. I began to sense the culture of the lab, the direction of the research, and the scale of what was possible in a system where science was not only imagined but actively built.

And then, in November 2019, she offered me the position.

That offer arrived during one of the most emotionally turbulent periods of my life–when my brother's suffering was worsening, when uncertainty felt like the air we breathed, and when I was still trying to understand how a world of advanced medical possibility could exist alongside such unbearable limitations. Yet there was a door opening, not through privilege, not through a formal pipeline, but through a simple chain of events: a mentor tagging my name, a scientist being willing to listen, and an opportunity shaped by work, writing, and readiness.

The joy of the offer was shadowed by my brother's diagnosis. The doctor spoke to him alone, leaving my brother Junior Bomolo and me outside. When he came out, he simply said, "I have cancer. There's nothing they can do." We were speechless. We returned home, watching him suffer.

On January 23, 2020, my brother passed away–the same month I had planned to travel to San Diego. I postponed my trip in grief.

Grief, Questions, and Resolve

As I prepared to leave Kinshasa the following month, my heart was heavy. Two men I loved–my father and brother–had died of liver disease. I asked myself, *Was it science that failed them, or their lack of access to science? Are there diseases for which the world truly has no answers? Or were the answers simply out of reach for us?*

I left for San Diego, not only to learn nanoengineering but to seek those invisible fixes–to find ways that science could offer answers regardless of geography, wealth, or circumstance. With grief behind me and questions ahead, I crossed borders once more–not just between countries, but between the known and the unseen.

CHAPTER 3

BECOMING A SCIENTIST OF THE WORLD

Introduction

I entered the world of nanotechnology not through textbooks, but through lived experiences–personal brushes with disease, scientific curiosity, and a deep desire to build solutions where few existed. From tuberculosis in the lungs to cancer deep in tissues, I learned how particles invisible to the eye could carry medicines to places human hands could not reach. That journey began in a small lab in South Africa and grew to span continents, technologies, and life-changing discoveries.

First Steps in Nanoparticle Research

I first started conducting research on nanoparticles for drug delivery at Rhodes University in South Africa, long before I ever set foot in San Diego. By way of introduction, nanoparticles are extremely small structures–typically between one and one thousand nanometers in size–with one nanometer being equal to one billionth of a meter. At such scales, materials can interact closely with biological molecules inside the body, because many cellular structures and biomolecules themselves operate at the nanoscale. This unique size range makes nanoparticles excellent vehicles for transporting drugs, vaccines, or diagnostic agents directly to specific sites within the body. In later chapters I will explain these concepts in simple terms, but at this stage of my career, I was simply fascinated by the possibility that something so tiny could solve such big problems.

My entry point into nanoparticle research was tuberculosis (TB). At the time, my postgraduate supervisor, Professor Rui Krause, was running a project exploring lipid-based nanocarriers for anti-TB drug delivery. This was before my brother fell ill and before cancer became a personal mission for me. In simple terms, lipid-based nanoparticles are microscopic "fat-based bubbles" engineered to carry a drug safely through the bloodstream and release it where it is most needed. My early work was driven by three intertwined motivations: a scientific challenge, a personal memory, and a sense of responsibility to address global health inequities.

The personal motivation dated back to 2002, when my father, stepmother, and I visited my aunt Germaine Nkanga in Kinshasa. During our stay, I became close friends with my cousin, Patou Iyongo, who fell gravely ill during the visit. We shared a room, meals, and many conversations. Toward the end of our stay, Patou was diagnosed with tuberculosis and passed away from the disease a few months later. My family panicked, worried that I had been exposed to the pathogen. My stepmother told my father one night, "That boy, Christian, must have caught that bug–he even ate food left by Patou." My father immediately sent me to TB specialists in Mbandaka for tests and follow-up. I was fortunate: No disease was detected, and I never developed TB. Years later, during my guest lectures for honors students at Rhodes University, I would joke that "tuberculosis fears me," but in truth, that episode stayed with me. It made TB more than just an abstract disease; it was part of my family's story.

Scientifically, TB fascinated me because of its stealth. The bacterium, *Mycobacterium tuberculosis,* hides deep inside lung tissues, often within macrophages, making it difficult for conventional drugs to reach and kill it. This is why TB treatment relies on multiple drugs administered over long periods–conditions that make drug resistance almost inevitable in many settings. The problem is especially acute in low- and middle-income countries (LMICs), where TB remains one of the most neglected diseases. I was driven by the idea that smarter drug carriers–nanoparticles–could help deliver existing drugs more efficiently to their hidden targets.

My early research focused on using crude soybean lecithin (CSL), a cheap and naturally occurring lipid material, to prepare liposomes, one of the most established classes of drug-delivery nanoparticles. The idea was simple but powerful: If we could use CSL to make liposomes that encapsulate TB drugs effectively, we could develop affordable nanomedicines suitable for resource-limited settings. I worked on formulating CSL-liposomes for isoniazid and rifampicin, two key TB drugs with challenging physicochemical properties. By carefully optimizing the formulation process, I managed to achieve encapsulation efficiencies as high as 80 percent for isoniazid–remarkable at the time for such a low-cost material.

These results opened doors. I presented my findings at the All-Africa Congress on Pharmacology and Pharmacy in Gauteng, South Africa, in October 2016. It was the first time I had presented my work to such a large and international scientific audience. Soon afterward, I published my first peer-reviewed paper in the *International Journal of Pharmaceutics*, titled "Preparation and characterization of isoniazid loaded liposomes made of crude soybean lecithin." This paper became a turning point in my scientific journey. Building on this initial work, I went on to conduct more advanced studies during my doctoral research–work I will explain later in the book in a more accessible way.

Building a Global Scientific Voice

Following this early success, opportunities came rapidly. I was invited to present at the Fifth International Symposium on Phospholipids in Heidelberg, Germany, organized by the Phospholipid Research Center. Phospholipids are the fatlike molecules that form the basic structure of every cell membrane–and the same materials scientists can engineer into drug-carrying nanoparticles, such as liposomes. For a young researcher from the Democratic Republic of the Congo, presenting my research at an international gathering in Germany was both humbling and empowering. I became a member of the Pharmaceutical Society of South Africa and the Phospholipid Research Center, which even granted me a travel fellowship.

But scientific breakthroughs are not just about presenting data–they are about refining ideas. While the CSL-liposomes were promising, they released nearly 40 percent of their drug content within thirty minutes. This burst release raised concerns about off-target distribution, since TB bacteria hide deep inside tissues. I needed a way to trigger drug release more precisely at the infection site.

With this goal, I pursued doctoral research funded by the South African National Research Foundation and Rhodes University. I began developing stimuli-responsive liposomes, modifying their molecular structure so that they would remain stable in circulation but release their drug cargo only in response to physiological cues, such as the acidic environment inside infected macrophages. This work was technically demanding, requiring me to learn advanced molecular chemistry techniques to graft antibiotics onto liposomal surfaces in controlled ways.

The effort paid off. I published several peer-reviewed papers in leading pharmaceutical and nanotechnology journals, culminating in a landmark publication in *Scientific Reports*, part of the Nature portfolio. This was a promise I had made to my supervisor early in my PhD: that I would publish at least one paper in a Nature journal. Achieving it was a deeply emotional moment for both of us, especially because the review process took nearly a year. For me, it meant that ideas born from a context of scarcity could stand on equal footing with the world's best.

These scientific contributions did not go unnoticed. I received the South African Chemical Institute (SACI) Postgraduate Medal as one of South Africa's top-five innovative young chemists, recognizing innovation, independence, and enterprise. I was also selected for the Novartis Next Generation Scientist Program in Basel, Switzerland–an intensive research internship that brought together top young scientists from around the world. At Novartis, I worked under Dr. Farshad Ramazani in the Parenteral, Topical and Ophthalmic Formulation Unit, learning how industry translates laboratory concepts into products for patients. This period expanded my perspective beyond academic research. I coauthored a paper in *Advanced Drug Delivery Reviews* and contributed as senior author to a book, edited by Dr. Ramazadi, on "clinical applications of long-acting injectables," leading a global team of scientists. These experiences were my first steps into industry and translational science, which became foundational later when I moved into biotechnology leadership roles.

By the time I completed my doctoral studies in April 2019, I had built not just a technical skill set, but a scientific identity rooted in innovation, adaptability, and global collaboration. My work on liposomal nanocarriers has taken me from Mbandaka to South Africa, Germany, Switzerland, and beyond. Yet, the most transformative stage of my journey was still ahead.

PhD graduation day at the Iconic 1820 Settlers' National Monument of Rhodes University in Makhanda, South Africa (2019). From left to right: Dr. Xavier Siwe Noundou, Mr. Alain M. Bapolisi, Christian I. Nkanga, Professor Rui W.M. Krause, Dr Pitchou B. Ngoy. Photo credit: Nnamdi I. Okafor

Postdoctoral Leap: A New Frontier

In February 2020, just weeks after the passing of my brother, I boarded a plane to California to begin a new chapter as a postdoctoral researcher at the University of California San Diego (UCSD), in the laboratory of Professor Nicole Steinmetz. This transition was both thrilling and emotionally heavy. While I carried the excitement of working in one of the world's leading nanotechnology labs, I also carried deep personal grief and unresolved questions. Two members of my family–my father and my brother–had died from liver-related diseases. As I crossed the ocean, I asked myself quietly, *Was it science that failed them, or was it access to science that failed us?*

The Steinmetz Lab specialized in bioengineering plant viruses as nanocarriers for drugs, vaccines, and imaging agents. Unlike my work in South Africa, where I built nanoparticles from scratch, here I learned to repurpose natural nanoscale objects–specifically plant viruses–as prefabricated scaffolds for biomedical use. These viruses are harmless to humans but structurally ideal for delivering therapeutic payloads.

I will return to the work on plant virus nanoparticles later in the book. For now, I want to briefly highlight a few key projects from my time at UCSD, as I introduce my postdoctoral experience.

Shortly after I arrived, the COVID-19 pandemic erupted globally. This crisis became both a challenge and an opportunity for our lab. I began working on a single-dose, long-acting COVID-19 vaccine based on the cowpea mosaic virus (CPMV), a plant virus extracted from black-eyed peas (*Vigna unguiculata*). CPMV was used to display pieces of the SARS-CoV-2 spike protein on its surface, thereby training the immune system to recognize and neutralize the real virus. My role was to design a hydrogel-based delivery platform that would slowly release the CPMV particles over time, mimicking the effect of multiple booster doses with just one injection.

This work was driven by urgency: Multiple-dose vaccination schedules were logistically difficult to implement at global scale, especially in resource-limited settings. A single-dose vaccine that maintained prolonged immunity could save countless lives. Our preclinical studies in mice showed promising results: Animals receiving the hydrogel-based single-dose vaccine maintained neutralizing antibody levels for five months–significantly longer than those receiving the standard multidose vaccine regimen. This work was later published in *Biomacromolecules*, a leading journal in the field.

At the same time, we applied similar slow-release hydrogel strategies to colon cancer immunotherapy. Colon cancer is the second leading cause of cancer-related deaths worldwide, largely due to tumor recurrence after surgery. Previous work in the lab had shown that CPMV triggers powerful innate immune responses that can suppress tumors, but frequent injections were required because the particles degrade quickly inside the body. We developed a hydrogel formulation that encapsulated CPMV and released it slowly over time. In animal studies, a single intraperitoneal injection of the hydrogel formulation prevented tumor regrowth for several weeks, outperforming multiple free CPMV injections. This was a striking example of how delivery technology–not just the therapeutic agent itself–can transform treatment outcomes.

I also contributed to research on tobacco mosaic virus (TMV) as a platform for cancer theranostics–the combination of therapy and diagnostics. By chemically modifying TMV particles to carry both an immunotherapeutic drug and a photothermal agent, we created a multifunctional nanoparticle capable of locally destroying melanoma tumors through heat and simultaneously stimulating systemic immune responses. In animal studies, this approach achieved three-fold higher survival rates compared to controls, while inducing strong tumor-specific T-cell responses. Part of these findings were published in *Nanomedicine: Nanotechnology, Biology, and Medicine*.

A Personal Turning Point

Amid these professional breakthroughs, I faced an unexpected personal challenge. While living in San Diego, I was diagnosed with a serious liver disease. Having already lost my father and brother to liver-related conditions, the diagnosis struck deep. I underwent a series of medical evaluations–imaging, blood analyses, and specialist consultations. There were moments of fear, uncertainty, and reflection. But unlike in Kinshasa years earlier, here, solutions were within reach. The standard therapy for my condition involved high drug doses with potential long-term toxicity. However, thanks to advances in targeted delivery, a specialized therapy was available that required less than 10 percent of the standard dose, drastically enhancing efficacy while reducing the risk of future toxicity. I received this treatment, recovered fully, and returned to my research with renewed determination.

This personal experience crystallized for me the real-world power of nanomedicine. It was no longer just about formulating particles in a lab or publishing papers–it was about survival, access, and hope. I had experienced firsthand how living in one place versus another can determine whether a life-threatening disease is a death sentence or a treatable condition. The contrast between my brother's fate and mine was impossible to ignore.

This period at UCSD was the most intense blend of science and humanity I had ever experienced. On one hand, I was coauthoring patents, publishing in high-impact journals, and contributing to cutting-edge therapeutic platforms. On the other, I was a patient–

grateful for innovations born of decades of scientific work, and painfully aware of the inequities that determine who benefits from those innovations.

Looking Back: Invisible Answers

When I reflect on this chapter of my life, I see three parallel journeys converging:

- The scientific journey, from designing cost-effective TB nanocarriers in South Africa to engineering plant virus-based immunotherapies in the United States.

- The personal journey, shaped by family tragedies, resilience, and ultimately survival.

- The global journey, connecting laboratories, universities, and ideas across continents.

Each step reinforced my belief that invisible fixes–whether in the form of nanoparticles, networks of scientists, or the circulation of knowledge–hold the key to bridging scientific and geographical divides. These technologies are invisible to the naked eye, but their impact on lives can be profound.

As I stood in the UCSD laboratory, working with viruses too small to see yet powerful enough to change global health, I thought back to my childhood in Mbandaka. Back then, diseases felt like mysterious forces. Now, I was part of a community unraveling those mysteries and building tangible solutions.

Bridge to Part II

Before we explore what nanoparticles are, how they work, and why they are transforming modern medicine, it's important to understand *why* they matter–not just scientifically, but personally and globally. In the next chapters, I will take you inside the world of nanotechnologies for biomedicine, explaining complex ideas in accessible language, and showing how these tiny structures are shaping therapies for some of humanity's greatest health challenges.

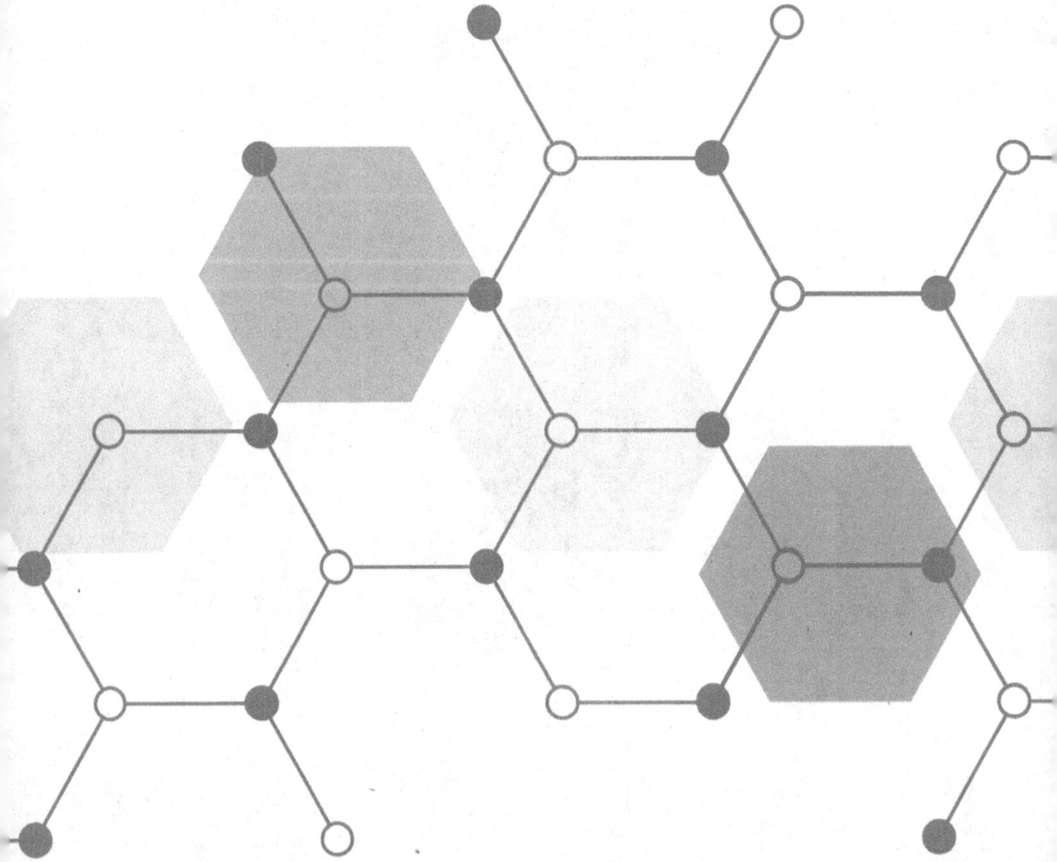

PART II

THE SCIENCE:

Nanotechnology & Healing

CHAPTER 4

INVISIBLE GIANTS:

The Power of Nanoparticles

Entering the Invisible World

Imagine holding something that looks empty but actually contains millions of tools powerful enough to fight cancer or infection. That is the world of nanotechnology. At this scale—one thousand times smaller than a strand of hair—matter behaves differently, and medicines can be guided with unprecedented precision. This chapter opens the door to that invisible world, showing how tiny particles can carry big solutions.

If you had asked my teenage self what the universe was made of, I would have answered with the objects I could see or touch: rocks, rivers, insects, bicycles, my school notebooks, the sky. The idea that the world could be shaped by things too small to see—forces too tiny to grasp, structures too minuscule to imagine—felt abstract, almost unreal.

And yet, during some of my high school breaks, I unknowingly performed my first "nanoscience experiment." I would tear paper into tiny scraps, place them on a bench, rub my pen on my hair, and bring it close. To my delight, the paper pieces jumped–leaping and dancing, as if obeying an invisible conductor. At the time, I did not know that this was a child's encounter with electrostatic forces, the same forces that move electrons, shape molecules, and hold atoms together. But that small moment planted a seed: The most powerful phenomena, the ones that shape life and determine health, are often the ones we cannot see.

Nanotechnology belongs precisely to that realm of the invisible.

To enter this world, imagine taking a grain of sand and dividing it into a billion pieces. Or think of a tennis ball. If the Earth were a tennis ball, then a nanometer would be something smaller than that tennis ball by the same unimaginable scale–far too small for the naked eye, yet powerful enough to control how molecules behave. Or imagine dividing a five-centimeter cube into nanometer-sized cubes; those cubes, spread in one layer, could cover an entire football field. Even more extraordinary, imagine a chunk of shiny yellow gold. Reduce it to nanoparticles, and it turns black or red or purple. Matter itself changes character. Color is no longer color, hardness no longer hardness. At the nanoscale, atoms rearrange, electrons shift, and surfaces dominate. Reality transforms.

This is the world of nanotechnology–where we build tools by manipulating matter at the level of atoms and molecules; where small particles become "invisible giants," capable of moving through the human body with precision; where biology and engineering meet in ways that allow us to imagine new medicines, new diagnostics, and new ways of understanding disease.

When Medicine Meets the Language of Life

Nanoparticles are broadly defined as structures between one and one thousand nanometers. Some are so small that they approach the size of DNA strands; others are about the size of viruses. But to understand why they matter, one must appreciate this simple truth: Life itself operates at the nanoscale. Proteins fold based on nanometer interactions. Hormones bind to receptors measured in nanometers. Viruses infect cells because their spikes are precisely arranged at the nanoscale. DNA stores life's code in nanometer-thick helices.

In other words, the nanoscale is not a scientific trick–it is biology's natural language. Nanoparticles are powerful because they can speak that language fluently.

Most medicines we use today, even the most sophisticated ones, are blind travelers. Take a pill, and it dissolves in the stomach, enters the bloodstream, and circulates everywhere–brain, kidneys, heart, gut, skin–hoping by statistical luck to reach the organ in need. As Sonia Contera writes in *Nano Comes to Life* (Princeton University Press, 2019), "Even when a drug is effective, it is very difficult to bring it to the right place." And this difficulty explains why drugs can be both miracle and menace: A cancer drug that kills tumor cells may also injure healthy cells; an antibiotic designed to save lives may harm the kidneys; a powerful medicine might work beautifully in a petri dish but fail inside the human body because it cannot reach its target. Nanotechnology tries to solve that, not by inventing magic, but by improving navigation.

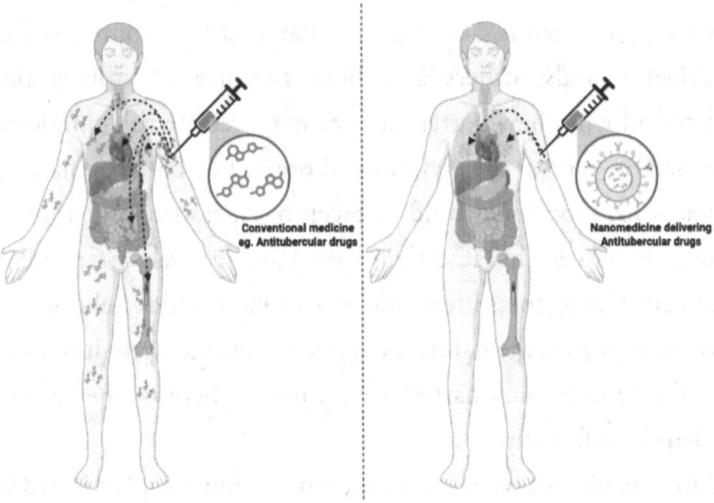

This figure illustrates the difference between conventional and nanoparticle-mediated delivery of antitubercular drugs. On the left, non-encapsulated (free) drugs are distributed throughout the body, exposing multiple organs to the medication and increasing the risk of systemic side effects. On the right, drugs loaded into nanoparticles are preferentially delivered to and accumulated in the lungs–the primary site of tuberculosis infection–enabling more targeted therapy, improved efficacy, and reduced off-target toxicity. (Courtesy of Mr. Pathy B. Lokole, who prepared this figure using BioRender). The image reflects a central idea of Invisible Fixes: even though these treatments are too small to see, they are carefully engineered to navigate the body's natural defenses and deliver drugs with precision.

Designing Precision: How Nanoparticles Work

A nanoparticle can be designed like a tiny courier–protected against the bloodstream, shielded from premature breakdown, and decorated with molecular "addresses" that guide it toward diseased cells. Some nanoparticles travel inside immune cells; some evade them. Some carry drug molecules deep inside their core; others display therapeutic agents on their surfaces. Some are activated by light, others by heat, others by acidity. Some are wholly synthetic; others are borrowed directly from nature.

There are two broad ways of building nanoparticles. The first is the top-down approach, which is similar to carving. Start with a large material and break it into smaller pieces using grinding, milling, or cutting. No chemistry changes–just the size. This is how flour is made from grains, or how stone is shaped into powder. This approach is robust but limited in precision.

The second approach, the bottom-up approach, is like growing crystals or building with Lego bricks. Molecules and atoms assemble into larger structures through predictable interactions–hydrogen bonds, van der Waals forces, ionic attractions, hydrophobic clustering. This method allows exquisite control over shape, size, and chemistry. Self-assembly, nanoprecipitation, sol-gel chemistry, and emulsification are common strategies. You choose the molecules; you define the environment; the particles grow according to the rules you set.

Bottom-up methods made modern nanomedicine possible.

But nature, far older than our laboratories, has long mastered nanotechnology. Viruses are naturally formed nanoparticles, exquisitely structured with precision far beyond our manufacturing capabilities. Exosomes–tiny vesicles released by cells–transport RNA and proteins between tissues. Even plants contain silica nanoparticles in their stems and leaves. Scientists now harvest and repurpose these natural nanostructures as delivery vehicles, imaging agents, or immune modulators.

Regardless of their origin, the effectiveness of nanoparticles usually depends on their composition. Metal nanoparticles made of silver, gold, copper, or zinc can have intrinsic antimicrobial activity.

Polymeric nanoparticles can dissolve slowly over time, releasing drugs in a controlled fashion. Lipid nanoparticles, like those used in mRNA vaccines, are soft, flexible, and biocompatible–perfect for carrying delicate molecules. Silica and porous materials can load large amounts of therapeutic compounds. Magnetic nanoparticles can be moved with external fields.

Some nanoparticles act as drug carriers; others act as drugs themselves.

One of the biggest challenges in designing nanoparticles is navigating the body's defense system. The immune system is vigilant and unforgiving. It recognizes foreign objects–bacteria, viruses, synthetic particles–and clears them out rapidly. The reticuloendothelial system (RES), composed largely of the liver, spleen, and macrophages, acts like a police network. Nanoparticles that appear "suspicious" are tagged by proteins in the blood (opsonization) and then engulfed by immune cells. This is helpful when the goal is to stimulate immunity, but unhelpful when you want nanoparticles to reach a microbe or tumor deep inside the body.

The solution is to disguise nanoparticles, so they appear native. Scientists often coat them with polymer chains like polyethylene glycol (PEG), or with natural proteins such as albumin. This disguise delays immune recognition, allowing nanoparticles to remain in circulation longer–sometimes hours, sometimes days–enough time to find their destination. Think of it, like giving a delivery truck the right license plates and permits so it can drive through checkpoints without being stopped.

Once in circulation, nanoparticles rely on two targeting strategies. The first is passive targeting, which takes advantage of natural leaks in diseased tissues. Tumors, for example, have chaotic blood vessels that are porous and poorly regulated. Nanoparticles can slip through these gaps and accumulate inside–this phenomenon is known as the *Enhanced Permeability and Retention* (EPR) effect. It is not perfect–EPR varies widely between tumor types and between patients–but it is a signature advantage of nanoscale drug carriers.

The second strategy is active targeting, in which nanoparticles are equipped with ligands–antibodies, peptides, sugars, or small molecules–that can bind to specific receptors on diseased cells. These ligands act as biological "addresses," helping nanoparticles dock at precise locations. Once docked, nanoparticles can release their cargo either gradually or in response to specific signals.

Controlled release is the next frontier of precision. Instead of releasing drugs immediately, nanoparticles can be engineered to respond to internal cues such as acidity (often elevated in tumors), enzyme activity (inflammation), or oxidative stress. External triggers such as light, heat, ultrasound, or magnetic fields can also activate release. This is what makes nanomedicine so unique: it can deliver the right dose at the right place at the right time.

As I progressed through my training, I began to truly appreciate the power of these invisible structures the first time I watched a batch of nanoparticles forming under controlled conditions in the lab. At Rhodes University, during my earliest experiments, I vividly remember preparing a solution that looked perfectly clear–nothing that the naked eye could distinguish from pure water. Yet, under the right illumination and after running the mixture through a particle size analyzer (*Zetasizer*), I saw the unmistakable signature peaks that revealed the presence of beautifully formed nanoparticles. It was humbling: A beaker that looked empty to the human eye actually contained millions of engineered structures, each one capable of carrying medicines or interacting with cells in ways the bulk drug never could. Standing there with my notebook, watching abstract theory turn into a tangible physical phenomenon, I understood for the first time that nanotechnology was not just a scientific concept. It was a toolkit–a new set of instruments, smaller than imagination yet powerful enough to reshape how we treat disease. That moment changed the way I saw matter. It was a reminder that what we call "empty" is often full, and that life's most transformative solutions often reside in realms too small for our senses.

But the field is not without hurdles. A nanoparticle designed to circulate for a long time may struggle to bind effectively to its target. A nanoparticle engineered for strong binding may release its drug prematurely. Making particles smaller may reduce immune clearance but increase kidney filtration. Adding more surface protection to avoid the immune system may improve circulation time but produce steric interference when it comes to binding to the target. Nanotoxicology, therefore, is not peripheral–it is central. Researchers must assess how nanoparticles interact with cells, tissues, organs, and biochemical pathways. They must test for inflammation, oxidative stress, DNA damage, organ accumulation and functions, and long-term clearance.

The example of doxorubicin illustrates this delicate balance. Doxorubicin given on its own (not protected inside a carrier) is potent against cancer but can damage heart tissue, if not dosed properly. When doxorubicin is encapsulated in liposomes, its toxicity drops dramatically–because the liposome shields the drug until it reaches the tumor. Yet some liposomes are inevitably trapped in immune organs, requiring careful attention to dose, composition, and clearance.

Despite these complexities, nanomedicine has moved far beyond theory. It has entered hospitals, clinics, and pharmacies worldwide through dozens of approved products. Liposomal formulations of chemotherapies reduce toxicity. Polymeric nanoparticles extend drug release. Iron oxide particles improve imaging. Lipid nanoparticles deliver mRNA vaccines that reshaped the COVID-19 pandemic response. Antibody-nanoparticle conjugates are emerging as new precision tools. And the field continues to grow–quietly, steadily, like a revolution that operates below the limits of human vision.

Nanotechnology is not magic. It is engineering guided by biology, biology refined by chemistry, and chemistry informed by physics. It is the art of making the invisible visible; of designing tools so small that they can slip into the inner architecture of the body and help solve some of the most stubborn problems in medicine.

In the next chapter, we will turn from principles to reality. We will examine some of the nanomedicines already approved for clinical use–stories of technologies that began as laboratory curiosities and are now transforming patient lives around the world. They demonstrate that nanotechnology is not just a promise; it is already a practice. And they show that the smallest tools can sometimes address the biggest challenges.

CHAPTER 5

NANOMEDICINES IN ACTION:

When the Invisible Becomes Clinical

Introduction

Many of the medicines discussed in this chapter do not introduce new active ingredients. Instead, they redefine how those ingredients behave once they enter the human body. By altering distribution, exposure, persistence, and tissue selectivity, nanotechnology has quietly changed the clinical fate of drugs that were once limited by toxicity, instability, or poor targeting. The nanomedicines presented here represent concrete evidence that formulation science can be as transformative as molecular discovery itself. Each example illustrates how nanoscale design decisions–often invisible to patients–translate into measurable differences in safety, tolerability, dosing strategy, and therapeutic reach. Together, they show that some of the most consequential advances in modern medicine are not loud breakthroughs, but subtle redesigns that reshape outcomes while remaining largely unseen.

For much of the twentieth century, the idea that microscopic engineered particles could safely circulate through the human body, locate diseased tissue, and release medicine precisely where it was needed seemed more imaginative than realistic. Yet, in a remarkably quiet scientific transformation, that vision has become part of everyday clinical practice. Nanotechnology applied to medicine–commonly referred to as nanomedicine–has matured from an experimental curiosity into a clinically validated discipline that now delivers approved therapies, improves patient outcomes, and reshapes treatment paradigms across multiple disease areas.

This evolution did not occur overnight. Over several decades, researchers systematically learned how materials behave at the nanoscale, how biological systems interact with particles of similar size, and how careful engineering could convert these interactions into therapeutic advantage. Many early concepts failed. Others required years–sometimes decades–of refinement before reaching patients. But today, the field has crossed a critical threshold: Nanomedicine is no longer experimental. It has successfully navigated the most difficult transition in biomedical innovation–the journey from laboratory discovery to regulatory approval and routine clinical use. In modern hospitals, infusion centers, specialty pharmacies, oncology clinics, and respiratory units, nanomedicine is no longer an idea. It is a set of products with barcodes, protocols, contraindications, monitoring parameters, and measurable outcomes in real patients.

Nanomedicine, at its core, is not merely a "smaller version" of traditional drug formulations. It represents a pharmaceutical redesign–one that fundamentally alters how a molecule behaves in the body.

Today, nanomedicines already serve millions of patients worldwide. Some address longstanding pharmaceutical challenges such as poor solubility, instability in the bloodstream, or rapid systemic clearance. Others protect healthy tissues from toxic exposure, enabling higher or more sustained dosing where conventional formulations would fail. Many exploit tumor biology, immune pathways, or microbial niches to concentrate drugs where they are most needed, while sparing the rest of the body. Collectively, these products demonstrate that nanotechnology does not merely enhance existing medicines–it often redefines how those medicines function clinically.

This is why nanomedicines matter to a broad healthcare audience. For physicians, they introduce new therapeutic options, new dosing strategies, and different risk/benefit tradeoffs. For nurses, they reshape infusion practices, monitoring priorities, and patient education. For pharmacists, they impose formulation-specific requirements for preparation, handling, substitution, and medication-use policy that demand precision and vigilance.

The central practical lesson is simple but critical: A nanoparticulate product is not automatically interchangeable with the "same drug" in conventional form. Even when the active ingredient carries the same name, its clinical behavior can be profoundly different–sometimes safer, sometimes riskier, but always formulation dependent.

Each example presented in this chapter illustrates a different way nanotechnology reshapes medicine–by improving efficacy, reducing toxicity, extending therapeutic action, or enabling entirely new treatment strategies.

In the following pages, I explore selected, clinically established nanomedicines. The aim is not to catalogue every approved nano-product, but to illustrate how and why these formulations succeeded: the problems they solved, the benefits they introduced, the limitations that remain, and the considerations clinicians must keep in mind at the bedside.

The information presented in this chapter is synthesized from public regulatory summaries, manufacturer disclosures, clinical-trial reporting, and pharmacological principles, interpreted through the lens of pharmaceutical sciences and translational medicine.

Nanomedicines in Oncology

ABRAXANE®—
Paclitaxel reengineered for safer delivery and different exposure

Paclitaxel is a cornerstone of chemotherapy. Its problem has never been its potency–it has been its formulation. Paclitaxel is poorly soluble in water, and earlier commercial products relied on solvent systems (notably Cremophor EL) to make infusion feasible. Those solvents are not inert; they can trigger hypersensitivity reactions and contribute to dose-limiting toxicity and infusion complexity.

ABRAXANE® reworks paclitaxel into albumin-bound nanoparticles–turning a solubility obstacle into a delivery strategy. Albumin is a naturally abundant carrier protein in plasma, optimized by physiology to transport hydrophobic molecules. In ABRAXANE®, paclitaxel associates with albumin through reversible interactions, forming particles on the order of a few hundred nanometers–small enough to behave as colloids in circulation, yet large enough to avoid immediate renal filtration.

What changes clinically with the nanoparticle version?

1. **Excipients change, and that changes safety.**
   Removing problematic solubilizers reduces infusion-related hypersensitivity risk. This often simplifies

premedication strategy and shortens infusion time, improving operational flow in infusion centers.

2. **Exposure patterns change.**
   Nanoparticle association can modify paclitaxel's apparent distribution and clearance, often producing different peak levels and tissue penetration than solvent-based paclitaxel. For clinicians, this matters because efficacy and toxicity are not solely dictated by the total dose–the time course of exposure is critical.

3. **Dosing feasibility changes.**
   Because dose-limiting toxicity can be partly formulation-driven, some patients may tolerate clinically effective exposure with fewer solvent-associated complications.

What remains essential for clinicians and nurses?

Even with improved tolerability, paclitaxel remains a cytotoxic microtubule inhibitor. Clinicians must still manage the foundational paclitaxel risks: neuropathy, myelosuppression, and cumulative toxicity. The nanoparticle does not remove the need for vigilance–it changes where vigilance is most valuable.

Pharmacy practice note:

Nanoparticle formulations should be treated as distinct products, not "generic paclitaxel." Policies on substitution must be explicit because formulation determines both tolerability and clinical handling.

DaunoXome®—
A Liposomal Anthracycline: Controlled Distribution Without Maximum Stealth

Anthracyclines such as daunorubicin are powerful anticancer agents with well-known systemic toxicities. DaunoXome® represents an early-generation liposomal strategy designed to modify the pharmacokinetics and tissue distribution of daunorubicin without fully extending circulation time. In DaunoXome®, daunorubicin is encapsulated within small non-PEGylated liposomes composed of distearoylphosphatidylcholine and cholesterol.

Unlike some later "stealth" liposomal systems, DaunoXome® is not designed to maximize circulation time indefinitely. Instead, it sits in a middle space: extended relative to free drug, yet not as prolonged as PEGylated liposomes.

Why does that matter?

Longer circulation can increase tumor exposure via passive accumulation, but it can also increase exposure in tissues where prolonged presence is undesirable. A non-PEGylated liposome may reduce certain skin toxicities associated with long-circulating systems, while still delivering a clinically meaningful modification of PK and tolerability.

Clinical framing:

For Kaposi's sarcoma and selected trial settings, DaunoXome® illustrates a key point: nanomedicine is not always about maximizing stealth. It is about optimizing the balance between:

- Exposure where you want it
- Limited exposure where you don't
- Predictable administration
- Acceptable cumulative toxicity

Nursing note:
Liposomal products can have different infusion reaction profiles and monitoring needs than free drugs–protocols should follow product-specific guidance rather than class assumptions.

DOXIL® / CAELYX®— A Turning Point in Cancer-Supportive Safety: Reducing Cardiotoxic Exposure

As the first FDA-approved cancer nanomedicine, DOXIL® is often described as historically important–and it is. But clinically, its true achievement is more specific: It changed the relationship between doxorubicin and the heart. By encapsulating doxorubicin within PEGylated liposomes, this formulation fundamentally altered the drug's behavior in the body.

Free doxorubicin distributes rapidly, including into myocardial tissue. Cumulative exposure can lead to cardiomyopathy, a risk that becomes more pressing as patients live longer or require multiple lines of therapy. In DOXIL®, doxorubicin is encapsulated within PEGylated liposomes–long-circulating carriers that reduce the amount of free drug immediately available to sensitive tissues like the heart.

What is the real-world benefit of this nanoparticle?

1. **Organ protection through confinement**
   By keeping most of the drug inside the liposome during early circulation, the formulation reduces exposure to tissues that accumulate free doxorubicin.

2. **Tumor selectivity by physiology**
   Many tumors have leaky vasculature and impaired lymphatic drainage. As explained in chapter 4, long-circulating nanoparticles can accumulate passively in tumor tissue (EPR effect). This does not mean "perfect targeting," but it can shift the distribution balance.

3. **Therapeutic feasibility**
   DOXIL® may enable continued use of anthracycline therapy in clinical situations where toxicity risk would otherwise limit options.

Important clinical caution:
Prolonged circulation is not "free." It can be associated with specific adverse events (e.g., skin and mucosal toxicities in some liposomal systems). The formulation changes the toxicity map; it does not erase toxicity.

Pharmacy and nursing note:
Because liposomal anthracyclines can have different infusion-related considerations and toxicity profiles than conventional doxorubicin, they should be managed as distinct agents in order sets and counseling materials. Interchangeability assumptions are unsafe.

FYARRO®— Sirolimus Delivered Intravenously via Albumin Nanoparticles for Rare Tumors

Sirolimus (rapamycin) is a potent mTOR pathway inhibitor. Its challenge, in many contexts, is not that it lacks biological effect, but that oral delivery may not reliably achieve the desired exposure at the tumor site due to absorption variability, metabolism, and systemic tolerability constraints.

FYARRO® introduces sirolimus as an albumin-bound nanoparticle administered intravenously. The intent is straightforward: bypass gastrointestinal limitations and deliver the drug in a form designed to achieve stronger and more consistent intratumoral exposure.

Why the nanoparticle matters here:

- **Exposure control:** IV administration plus nanoparticle association can yield more predictable systemic delivery than oral dosing.
- **Tumor pharmacodynamics:** improved local exposure can translate into more sustained pathway inhibition where it matters clinically.
- **Feasibility in rare cancers:** for diseases with limited therapeutic options, incremental improvements in drug delivery can become clinically decisive.

For clinicians, FYARRO® is also a reminder that nanomedicine is not only about "new drugs." It is often about making known drugs clinically usable in contexts where conventional delivery fails.

MARQIBO®—
Liposomal Vincristine and the Complex Reality of Post-Approval Evidence

Vincristine is effective, but neurotoxicity can limit its dosing. Liposomal encapsulation aims to change vincristine distribution and exposure kinetics so that effective antitumor dosing becomes more feasible.

MARQIBO® offers an important teaching case for clinicians and trainees: Accelerated approvals can provide earlier access for high-need diseases, but they are coupled to requirements for confirmatory evidence. When confirmatory trials are delayed or incomplete–sometimes due to recruitment challenges rather than scientific failure–regulatory status can change.

Clinical lesson:

Nanomedicine lives in the real world, where science, ethics, feasibility, and logistics intersect. Evidence generation is part of the product lifecycle, not a one-time event.

ONCASPAR®—
PEGylation as a "Nanomedicine-Adjacent" Strategy That Changes Half-Life and Immunogenicity

Not all clinically important nanoscale strategies involve encapsulating a drug inside a carrier. PEGylation is one of the clearest examples:

covalently attaching polyethylene glycol to a therapeutic enzyme increases its apparent size and alters its interaction with the immune system and clearance pathways.

Pegaspargase (ONCASPAR®) extends enzyme activity duration and reduces dosing frequency compared with native asparaginase. For pediatric oncology teams, the practical benefit is substantial: Fewer administrations can mean less procedure-related anxiety, fewer clinic events, and more consistent therapeutic depletion of the target amino acid.

Clinical value proposition:

- **Less frequent dosing** (time and burden reduction)
- **More sustained pharmacologic activity** (therapeutic continuity)
- **Lower hypersensitivity risk** relative to some native preparations (though not eliminated)

Practice note:

PEGylation reduces–but does not eliminate–immunogenicity. Teams must still be prepared for allergic reactions, pancreatitis risk, hepatic effects, and thrombosis/bleeding complications consistent with asparaginase-class pharmacology.

ONIVYDE®—
Liposomal Irinotecan as a Strategy for Better Tolerability and Sustained Exposure

Irinotecan is a potent agent whose clinical use can be constrained by tolerability and by pharmacokinetic variability. Liposomal irinotecan aims to stabilize exposure, protect the payload from premature metabolism, and deliver a more sustained availability of active drug at relevant sites.

For metastatic pancreatic cancer–a disease in which incremental survival benefits matter deeply–liposomal delivery offers a practical advantage: It can make irinotecan therapy more clinically manageable in combination regimens by shifting exposure kinetics and improving the balance between efficacy and tolerability.

For clinicians:
The formulation does not eliminate irinotecan-associated adverse effects (e.g., GI toxicity). But by moderating the exposure profile, it may reduce intensity or frequency in some settings and allow integration into regimens that patients can sustain.

For pharmacists:
Liposomal irinotecan must not be treated as a simple substitute for conventional irinotecan. Product-specific preparation, dosing, and toxicity expectations apply.

Nanomedicines in Infectious Diseases

AmBisome®—
A Classic Example of Nanomedicine as Toxicity Management

Amphotericin B is among the most effective broad-spectrum antifungals ever used–yet its conventional formulation is notoriously nephrotoxic. AmBisome® encapsulates amphotericin B within liposomes, reducing free-drug exposure to renal tissue while preserving antifungal activity.

This is nanomedicine at its most clinically intuitive: same potent drug, safer delivery.

Why it matters at the bedside:

- For febrile neutropenia and severe systemic fungal infections, clinicians often need rapid antifungal therapy that can be sustained.

- Reduced nephrotoxicity can allow therapy continuation in patients who would otherwise require dose reduction or discontinuation.

- Nursing teams see the benefit as fewer infusion-limiting adverse events and fewer therapy interruptions due to renal function decline.

Critical pharmacy point:

Different amphotericin products are not interchangeable on a milligram-per-milligram basis. Lipid-based, liposomal, and conventional formulations have different distribution, tolerability, and dosing logic. Institutional protocols should explicitly prevent inadvertent substitution.

ARIKAYCE®— Lung-Targeted Nanomedicine: High Local Exposure, Low Systemic Burden

Mycobacterial lung disease is not only a microbiology problem; it is a delivery problem. The organisms can reside in biofilms and within macrophages, making systemic antibiotic exposure insufficient or poorly tolerated.

ARIKAYCE® delivers amikacin encapsulated in liposomes via inhalation, concentrating exposure in the lungs and within macrophages while keeping serum levels relatively low compared with IV dosing.

Clinical meaning:

- Higher local drug concentrations where the organism resides
- Enhanced intracellular delivery to macrophages
- Reduced systemic toxicity risk relative to high-dose systemic aminoglycosides

Nursing and respiratory therapy note:
Because administration depends on a specific nebulization system, adherence and technique are part of efficacy. Patient education

(device use, cleaning, consistent dosing routine) becomes therapeutic–not optional.

Nanomedicines Beyond Oncology and Infection

COPAXONE®— A "Non-Biologic Complex Drug" Where Structure Is the Therapy

COPAXONE® is a reminder that not all nanomedicines look like classic nanoparticles. Glatiramer acetate is a complex mixture of polypeptides that self-associate in solution. Its biological activity emerges from its complex composition and nanoscale organization rather than a single defined molecular entity.

For clinicians, the main point is practical: This is a therapeutically established immunomodulator with robust outcomes in relapsing multiple sclerosis. For scientists, it is a conceptual bridge: nanostructure itself can be the medicine, not merely a carrier.

Practice note for teams:

The dosing schedule (daily or three times weekly, depending on the product format) has direct implications for adherence, injection-site reactions, and patient confidence–areas where nursing support and counseling make a measurable difference.

INVEGA SUSTENNA® / TRINZA®— Nanocrystal Depots ("Medicine Reserves") and the Clinical Importance of Adherence

In psychiatry, adherence is not a "soft" endpoint; it is often the dividing line between stability and relapse. Long-acting injectable antipsychotics exist because real life is complex: patients forget, disengage, struggle with side effects, lose access, or simply cannot maintain daily oral routines during unstable periods.

Paliperidone palmitate nanocrystal suspensions are engineered so that slow dissolution and depot kinetics maintain therapeutic plasma levels across extended dosing intervals.

Why the nanocrystal matters:

- Predictable release through controlled dissolution
- Reduced need for daily adherence decisions
- Potential reduction in relapse driven by missed oral doses
- Simplified logistics for caregivers and care teams

Clinical nuance:

Depot therapies are not "set-and-forget." They require structured follow-up, monitoring for EPS/metabolic effects, and careful attention to transition strategies when switching products or restarting after missed doses.

Placed at the conclusion of this chapter, **Table 5.1** captures the essence of *Invisible Fixes* in clinical practice. Each nanomedicine summarized here operates at a scale unseen by the human eye, yet its impact on patient care is tangible–altering drug behavior, improving safety, extending efficacy, and redefining what is therapeutically possible. These formulations demonstrate that innovation does not always announce itself dramatically; often, it works silently through formulation science, particle engineering, and biological compatibility. At the same time, their existence exposes a deeper challenge: While these invisible fixes are scientifically mature and clinically validated, access to them remains uneven across health systems and geographies. Ensuring that such quiet breakthroughs benefit all patients is not only a scientific responsibility, but a moral and societal one.

Table 5.1. Selected clinically established nanomedicines and their key nanotechnological attributes

Product (Brand)	Indication(s)	Nanotechnology platform/ composition	Approx. particle size	Key nanotechnology attribute	Clinical advantage vs. conventional form
ABRAXANE® (paclitaxel albumin-bound)	Breast cancer, pancreatic cancer, NSCLC	Albumin-bound paclitaxel nanoparticles (protein-based nanosuspension)	~130 nm	Carrier-free solubilization via endogenous albumin	Eliminates Cremophor EL toxicity; higher MTD; shorter infusion time; improved tumor penetration
DaunoXome®	Kaposi's sarcoma	Non-PEGylated liposomal daunorubicin (DSPC/cholesterol)	~45 nm	Liposomal confinement alters biodistribution	Reduced systemic toxicity; intermediate circulation time; lower skin toxicity than PEGylated liposomes

DOXIL® / CAELYX®	Ovarian cancer, Kaposi's sarcoma, multiple myeloma	PEGylated liposomal doxorubicin (STEALTH® liposomes)	80–100 nm	Long-circulating stealth liposomes (PEGylation)	Markedly reduced cardiotoxicity; prolonged half-life (from < 10 min to ~74 h); enhanced tumor accumulation (EPR effect)
FYARRO® (nab-sirolimus)	Malignant PEComa	Albumin-bound sirolimus nanoparticles	~100 nm	Nanoparticle albumin-bound delivery	Improved intratumoral delivery; enhanced mTOR inhibition; superior pharma-codynamics vs oral sirolimus
MARQIBO®	Relapsed ALL (withdrawn)	Liposomal vincristine (sphin-gomyelin/choles-terol)	~100 nm	Liposomal encapsulation increases expo-sure	Enabled higher vincristine dosing; altered PK; approval with-drawn due to postmarketing trial issues

Product (Brand)	Indication(s)	Nanotechnology platform/ composition	Approx. particle size	Key nanotechnology attribute	Clinical advantage vs. conventional form
ONCASPAR® (pegaspargase)	Acute lymphoblastic leukemia	PEGylated enzyme (not encapsulated; nanoscale conju-gate)	<5 nm (hydrodynamic)	PEGylation extends circulation	Prolonged half-life; reduced immunogenicity; fewer injections vs native asparaginase
ONIVYDE®	Metastatic pancreatic cancer	PEGylated liposomal irinotecan	~110 nm	Controlled release + metabolic protection	Reduced toxicity; prolonged SN-38 exposure; improved overall survival and progression free survival vs standard regimens

AmBisome®	Systemic fungal infections, leishmaniasis	Liposomal amphotericin B (HSPC/DSPG/cholesterol)	~80–100 nm	Liposomal membrane shields toxic drug	Dramatically reduced nephrotoxicity; improved tolerability vs amphotericin B deoxycholate
ARIKAYCE®	MAC lung disease	Inhaled liposomal amikacin	~300 nm (aerosolized liposomes)	Targeted pulmonary delivery; macrophage uptake	High lung exposure; minimal systemic toxicity; biofilm penetration
COPAXONE®	Relapsing multiple sclerosis	Self-assembled polypeptide nanostructures (NBCD)	~1–1000 nm (colloidal range)	Intrinsically bioactive nanostructure	Immunomodulation without classical carrier; reduced relapse rates; long clinical history

Product (Brand)	Indication(s)	Nanotechnology platform/ composition	Approx. particle size	Key nanotechnology attribute	Clinical advantage vs. conventional form
INVEGA SUSTENNA®	Schizophrenia	Paliperidone palmitate nanocrystal suspension	D50 = 650 – 940 nm	Nanocrystal-controlled dissolution	Monthly dosing; no oral overlap; improved adherence
INVEGA TRINZA®	Schizophrenia	Nanocrystal LAI (paliperidone palmitate)	<1000 nm	Extended nanocrystal depot	3-month dosing interval; stable plasma levels; reduced relapse risk

SN-38 – 7-Ethyl-10-hydroxycamptothecin: *The active metabolite of irinotecan responsible for its antitumor activity. SN-38 is significantly more potent than the parent drug but is poorly soluble and rapidly cleared, which is why nanoparticulate formulations (such as liposomal irinotecan) are designed to optimize its delivery and exposure at tumor sites.*

What Health Professionals Should Take Forward

Across oncology, infection, neurology, and psychiatry, a consistent theme emerges: Nanomedicines succeed when they solve a clinically blocking problem that conventional formulations cannot reliably solve. The most common wins include:

1. **Reducing formulation-driven toxicity** (e.g., nephrotoxicity, hypersensitivity, organ exposure)
2. **Improving delivery to difficult sites** (tumor tissue, intracellular macrophage compartments, lungs)
3. **Stabilizing pharmacokinetics** (more consistent exposure, fewer peaks/troughs)
4. **Enabling practical care** (shorter infusions, less frequent dosing, easier regimen adherence)

But nanomedicine also introduces new responsibilities: product specificity, non-interchangeability, formulation-dependent adverse events, and the need for tighter coordination across the care team.

In the next chapter, we move from "nanomedicines already in the clinic" to a frontier that is increasingly shaping chronic disease management and global health equity: long-acting therapies and the engineering of time in medicine.

CHAPTER 6

ENGINEERING TIME:

Nanotechnology and the Rise of Long-Acting Medicines

Introduction

For millions living with chronic or life-threatening illnesses, the hardest part is not the medicine–it is the daily burden of taking it. Nanotechnology is changing that. By turning drugs into long-acting formulations that last weeks or months, scientists are rewriting what it means to treat disease. This chapter explores how these innovations reduce suffering, restore dignity, and challenge global health systems to deliver therapies that patients can truly live with.

Time is one of the most powerful forces in medicine. It governs how diseases progress, how drugs act, how healing unfolds, and how long a patient must maintain hope. Time decides whether a treatment feels manageable or unbearable. It shapes adherence, recovery, relapse, and, in many cases, survival. As a pharmacist and nanotechnology researcher, one lesson has followed me across every laboratory, hospital corridor, and patient story I've ever encountered: Medicine is never only about chemistry–it is about time, and how we engineer it.

Patients with chronic illnesses often describe their medication schedules as a second job. Pills in the morning, pills with meals, pills before sleeping. A patch on Tuesday, an injection on Thursday. Blood tests every month. Dose adjustments every few weeks. For people with cancer, HIV, schizophrenia, diabetes, rheumatoid arthritis, or chronic pain, "treatment" becomes a rhythm that dictates life. Some manage; many struggle. Not because they do not want to live–but because being sick already drains so much of their strength, leaving little room for complicated therapeutic regimens.

It is here, in this deeply human gap between biology and everyday life, that long-acting injectables (LAIs) emerged–not as scientific luxuries, but as compassionate technologies designed to give patients something almost priceless: the freedom to live without constantly thinking about medicine.

The concept sounds simple: instead of daily or weekly dosing, you give a single injection that slowly and safely releases the medication for weeks, months, or even a full year. A treatment becomes an event rather than a daily burden. Anxiety decreases. Adherence improves. Clinical outcomes rise.

But achieving this simplicity is anything but simple.

The Human Need That Preceded Technology

Before I ever studied long-acting medicines in a laboratory, I witnessed firsthand how much they could change a life.

Months before my older brother, Jean Corentin, passed away from liver cancer in 2020, his days were governed by tablets–some for pain, some for nausea, some simply to maintain minimal comfort. There were moments when he was too weak to swallow, and moments when the pills came back up as fast as they went down. And then moments when he looked at me with tears in his eyes and whispered, "Christian, give me a break." I could not. The disease would not.

My cousin, who passed away in 2025 after also battling an aggressive form of liver cancer, faced similar challenges. His medicines were not curative–only palliative. Yet even those palliative doses had to be taken regularly, on time, without interruption, or his pain returned with brutal force.

In these moments I understood something essential: Therapeutic failure often comes not because the drug is ineffective, but because the treatment schedule is inhumane.

A technology that transforms many daily doses into a single monthly or yearly intervention is not merely scientific innovation–it is the embodiment of empathy.

Why Not All Drugs Can Be Taken as Pills

Many people assume that medicines come as tablets because it is convenient. But biologically, the human body is a harsh and unforgiving environment for most drug molecules. Biomacromolecules–peptides, proteins, antibodies, RNA drugs–are too fragile. The stomach destroys them in minutes. The intestine blocks them. The liver metabolizes them before they can act. Even small molecules often suffer from poor solubility or permeability, making them incapable of entering the bloodstream in meaningful amounts.

Thus, many breakthrough therapies can only be delivered parenterally–through injections.

But injections introduce their own challenges. Once a drug enters the bloodstream, the body reacts like a strict gatekeeper. Enzymes degrade it. Immune cells engulf it. Kidneys filter it. Liver enzymes dismantle it. Some drugs vanish within minutes. That means more injections. More clinical visits. More pain. More anxiety.

This cycle prompted scientists to ask: Can we engineer a medicine that stays active for weeks or months without repeated dosing?

This question became the foundation of long-acting therapies.

Why Long-Acting Medicines Are Hard to Design

Designing a medicine that lasts a month in the body requires precise control over chemistry, physics, and biology. The drug must:

- Dissolve slowly but reliably
- Avoid crystallizing prematurely
- Resist enzymatic degradation
- Remain stable for months inside the body
- Be safe for long-term exposure
- Be sterile, manufacturable, reproducible
- Be scalable to millions of doses
- Degrade into harmless byproducts

This is why long-acting technologies often rely on complex materials such as biodegradable polymers, lipids, or crystalline matrices.

However, the biggest barrier is physical chemistry: *Most drugs (40 percent of approved molecules and 90 percent of drug candidates) are poorly soluble.*

If a drug does not dissolve well, it cannot form an injectable depot that releases slowly and predictably. Poor solubility means irregular release, dose dumping, or harmful local reactions. My own review with Dr. Farshad Ramazani in *Advanced Drug Delivery Reviews* revealed a striking fact: Only 8 percent of existing LAIs are made from soluble drug molecules. In essence, most drugs resist being turned into long-acting products. And here is where nanotechnology enters like a quiet but powerful hero.

How Nanotechnology Changed Long-Acting Delivery

Nanotechnology allows scientists to reshape a drug at the molecular and structural levels. By turning a drug into nanocrystals, embedding it into nanoparticles, or engineering smart nanoscale carriers, we can transform its solubility, stability, release rate, and distribution.

As discussed in chapter 4, some nanoparticles act like tiny vaults that leak medication slowly over weeks or months. Others respond to stimuli–pH changes, enzymes, oxidative stress–to release medication only when it reaches the right biological environment.

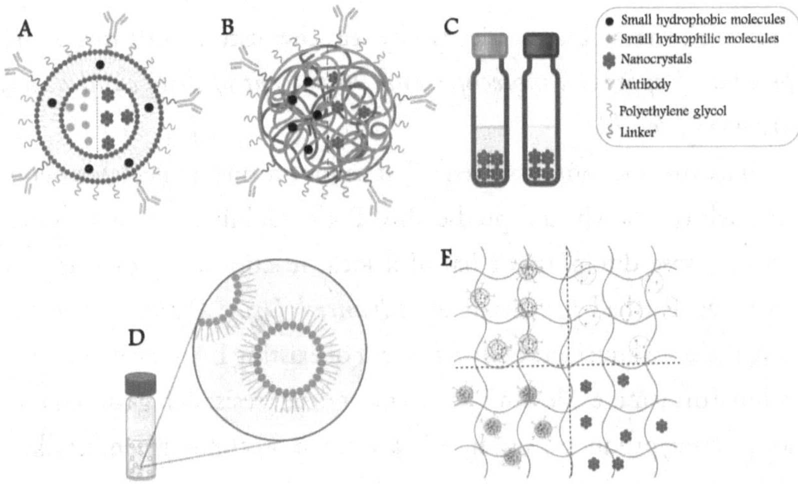

Overview of nanotechnology-enabled strategies for long-acting injectable (LAI) drug delivery (Courtesy of Mr. Pathy B. Lokole, who prepared this figure using BioRender). API stands for active pharmaceutical ingredient. This schematic illustrates several nano-based formulation platforms designed to enhance sustained release, improve drug solubility, and enable site-specific delivery:

A) Multifunctional liposomes–lipid vesicles capable of carrying drugs within their aqueous core or lipid bilayer, and adaptable with surface ligands for targeted and stimuli-responsive release. B) Polymeric nanoparticles–engineered particles that enable controlled, stimuli-triggered drug release through biodegradable or responsive polymer networks. C) Nanosuspensions of prodrug nanocrystals–poorly soluble or chemically modified drug crystals dispersed in (1) oily or (2) aqueous vehicles for slow dissolution and prolonged systemic exposure. D) Nanoemulsions–oil-in-water or water-in-oil droplets stabilized by surfactants, allowing solubilization and gradual release of lipophilic or hydrophilic drugs. E) Hydrogel-based depots incorporating nanosystems–injectable gels embedding liposomes, polymeric nanoparticles, nanogels, or nanocrystals to achieve long-term, localized release.

As illustrated in chapter 5, there are several nanotechnology-enabled LAIs currently used in the clinic. These include:

- **Liposomes** (DepoDur®, Myocet®)
- **Nanocrystals** (Invega Trinza®, Invega Hafyera®)
- **Polymeric nanoparticles** (Eligard®, Zilretta®)
- **Nanoemulsions** (Diprivan®)

These products show how nanotechnology can engineer time—not just drug action. But the innovation extends further.

Prodrug Nanocrystals:
Some long-acting medicines rely on a brilliant trick: attach a lipophilic chain to the drug, forming a "prodrug" that dissolves slowly over months. Once released into the bloodstream, enzymes cleave the chain and activate the drug. Examples for such products include Invega Sustenna® (a long-acting injectable antipsychotic medicine) and Zyprexa® (an oral antipsychotic drug with molecules intended to circulate freely in the bloodstream). Mainly used to treat schizophrenia (and related conditions such as schizoaffective disorder and bipolar disorder), these medicines transformed mental health care by providing stable blood levels and reducing relapse.

Semi-solid Prodrug Nanoparticles

A groundbreaking 2019 Nature Communications study developed semi-solid nanoparticles for HIV therapy–turning water-soluble antivirals into long-acting depots that lasted months. This approach could fundamentally change HIV care worldwide.

My Own Immersion and the Power of Slow-Release Immunotherapy

During my postdoctoral work at University of California San Diego (UCSD) in Professor Steinmetz's lab, I experienced firsthand how time engineering can change cancer therapy. We embedded plant-virus nanoparticles into hydrogels to create long-acting immunotherapies. In colon cancer recurrence models–a notorious clinical challenge–a single slow-release dose prevented tumor return entirely in mice. The moment we saw the results was unforgettable. A single dose had achieved what multiple injections could not.

And in that moment, I thought of my father and my brother. I wondered how differently their stories might have unfolded if therapies like these were accessible to them.

Nanotechnology had given me both a scientific revelation and a personal ache.

The Hidden Cost of Innovation: Manufacturing, Regulation, and Access

As elegant as nanotechnology is, long-acting formulations face hurdles far beyond the science:

- **Manufacturing is complex and expensive.** Nanoparticles require strict size control, sterile processing, stability testing, and advanced analytical methods.

- **Regulation has not caught up.**
  There are no universal guidelines for nanoparticle-based LAIs, causing uncertainty for pharmaceutical companies.

- **Stability is challenging.**
  Some nanoparticles aggregate, oxidize, or degrade over time. Lyophilization (freeze-drying) introduces new technical problems.

- **The immune system can clear nanoparticles prematurely.**
  Making long-acting depots requires overcoming several biological barriers.

- **Costs remain high.**
  Many patients cannot afford these technologies, especially in low-income settings.

This final point is deeply personal to me. If long-acting cancer immunotherapies had been available–and accessible–in Kinshasa or Mbandaka years ago, would my brother's story have been different? Would my father have lived longer? Would my cousin have had more than palliative care? I don't know. But I know that access should not depend on geography or socioeconomic status.

Why This Field Matters for the Future of Medicine

Long-acting nanomedicines hold immense potential to transform chronic disease management:

- Reducing pill burdens
- Improving adherence
- Minimizing side effects
- Lowering long-term healthcare costs
- Improving mental health
- Slowing disease progression
- Providing stability to families and caregivers

Beyond HIV, schizophrenia, pain, or hormonal therapy, long-acting nano-products can revolutionize:

- Cancer immunotherapy
- Tuberculosis treatment
- Autoimmune diseases
- Neurodegenerative disorders

- Reproductive health
- Pediatric chronic illness management

Imagine a world where a child with epilepsy receives one injection every six months; an HIV patient receives two injections per year; a cancer survivor gets a once-yearly immunotherapy boost; a diabetic patient receives quarterly metabolic stabilizers; women can rely on safe, reversible long-acting reproductive treatments; and people battling substance use disorder receive long-acting support without daily pills.

This is not a futuristic imagination. Science already exists. What is missing now is global commitment, investment, and equitable access.

A Technology of Empathy

If nanotechnology represents invisible precision, long-acting injectables represent invisible compassion. They reduce the daily noise of illness. They give back time. They restore autonomy. They simplify life. They preserve dignity. They soften suffering. For patients, this is not merely scientific convenience; it is emotional relief. It is hope. It is life lived with fewer interruptions.

In many ways, LAIs embody the central thesis of this book: that the most transformative solutions in medicine are often invisible—quiet technologies working behind the scenes to restore balance, stability, and freedom.

Closing Thought

As I think about the future, I see long-acting, nano-enabled therapies becoming central pillars of global healthcare. But I also see the danger of inequity. Technologies that could save millions may remain confined to high-income settings unless we act intentionally to ensure global impact of these products, to ensure all these innovative products reach all the patients in need. That is why this research is personal to me. Nanotechnology allowed me to engineer time in a laboratory–but life showed me why it matters.

My hope is that one day, no father, brother, daughter, or cousin will struggle to have access to long-acting, life-sustaining treatments simply because of where they were born. And perhaps, in that future, the invisible fixes we design today will become the visible healing the world urgently needs.

CHAPTER 7

GREEN NANOTECHNOLOGY:

Healing with What the World Throws Away

Introduction

Green nanotechnology introduces a different way of thinking about innovation at the nanoscale—one that begins not with advanced machinery, but with biological matter, local resources, and ecological logic.

By using plants, microorganisms, and organic waste as reducing and stabilizing agents, green nanotechnology demonstrates that nanoparticle fabrication does not have to depend on toxic chemicals, energy-intensive processes, or centralized industrial infrastructure. Instead, it leverages nature's intrinsic chemistry to generate functional nanomaterials whose properties are shaped as much by their biological origin as by their size.

What distinguishes this approach is not merely sustainability, but accessibility. Green nanotechnology aligns nanoscale science with real-world constraints, particularly in regions where advanced manufacturing capacity is limited. Agricultural by-products, plant extracts, and biological residues–often treated as waste–become starting materials for nanostructures with antimicrobial, therapeutic, environmental, and diagnostic relevance. In this framework, innovation is no longer imported; it can be cultivated locally.

These biologically derived nanoparticles often carry multifunctional characteristics. The organic molecules associated with their formation can influence stability, activity, and biological interaction, sometimes conferring properties that are absent in conventionally synthesized counterparts. Green nanotechnology thus reframes nanoparticles not as inert carriers or isolated materials, but as hybrid constructs shaped by both engineering intent and biological context.

In this chapter, green nanotechnology is presented not as an alternative curiosity, but as a complementary pathway–one that expands who can participate in nanoscience, how nanomaterials are produced, and where their benefits can reach. It exemplifies how invisible fixes can emerge from overlooked resources, transforming local matter into global opportunity.

Nanotechnology has already demonstrated its power to transform medicine. As illustrated in the preceding chapters, nanoscale engineering has enabled safer chemotherapies, longer-acting injectables, smarter vaccines, and targeted delivery systems that would have been unimaginable only a few decades ago. Yet these successes also reveal a troubling paradigm: While nanomedicines are increasingly embedded in advanced healthcare systems, their benefits remain largely inaccessible to much of the world.

In many low- and middle-income countries, including large parts of Africa, the barrier is not a lack of medical need or scientific talent–it is cost. Sophisticated infrastructure, energy-intensive manufacturing processes, specialized reagents, and capital-heavy industrial translation pipelines place conventional nanotechnology beyond the reach of many healthcare systems. As a result, some of the most powerful medical innovations ever developed remain concentrated in settings where resources are already abundant.

This imbalance raises a central question that lies at the heart of *Invisible Fixes*: Can nanotechnology be reimagined to serve global health equitably rather than deepen technological divides?

Green nanotechnology offers a compelling and practical response.

From Scarcity to Strategy

My understanding of this challenge did not emerge from theory alone. As a professor of pharmaceutical sciences at the University of Kinshasa, I confronted it directly through everyday academic and clinical realities. In 2021, within the framework of a research internship for a third-year student, Rebecca N. Monga, my team conducted a survey of thirty-four community pharmacies across Kinshasa, the capital of the Democratic Republic of the Congo. The findings were sobering: Only four pharmacies stocked any nanomedicine at all–and those products were limited to a small number of first-generation formulations.

If nanomedicines are scarce even in a capital city of more than fifteen million people, their absence in rural and underserved regions is almost guaranteed. This local observation reflects a broader global pattern: Nanotechnology has matured scientifically, but its translation into equitable access has lagged behind.

Green nanotechnology emerged in my work as a response to this gap. It is not merely an alternative synthesis route; it represents a conceptual shift. Instead of relying on expensive chemicals, high-energy equipment, and tightly controlled industrial environments, green nanotechnology leverages biological systems and naturally occurring materials to produce nanoparticles safely, affordably, and sustainably.

In essence, it asks a deceptively simple question: *What if the raw materials for advanced nanotechnology already exist in abundance–discarded, overlooked, or treated as waste?*

How Green Nanotechnology Works

At its core, green nanotechnology relies on a fundamental chemical principle: Metal ions can be reduced into elemental nanoparticles by electron-donating biological molecules. Plants, microorganisms, and biological residues contain a rich repertoire of compounds–such as polyphenols, flavonoids, proteins, enzymes, sugars–that can act simultaneously as reducing agents, stabilizers, and surface functionalizers.

When a metal salt such as silver nitrate is introduced into a biological extract, metal ions are reduced to their metallic state. These atoms nucleate and grow into nanoparticles, while surrounding biomolecules form a stabilizing organic layer that prevents aggregation and often enhances biological activity. Unlike conventional chemical or physical synthesis methods, this process avoids toxic solvents, operates under ambient conditions, requires minimal energy input, and produces biologically functionalized nanostructures by default.

The outcome is not merely a nanoparticle, but a bio-integrated nanosystem–one that is inherently compatible with living systems.

Why Plants–and Waste–Matter

Among biological platforms, plants and plant-derived waste have emerged as particularly powerful tools for green nanotechnology. Plant extracts act rapidly, reproducibly, and at scale. More importantly, they are widely accessible.

In many developing regions, agricultural waste accumulates with limited economic value. Banana peels, stems, leaves, and pseudostems are often discarded or used only as low-grade fertilizer. In French-speaking cultures, there is even a common expression– "*mettre la peau de banane*" –used to describe setting a trap for someone to fall or fail. The phrase originates from the familiar image of slipping on a banana peel.

That metaphor stayed with me.

What if what makes people fall could instead help the world stand?

From Banana Waste to Nanotechnology Platforms

When I learned about the African Research Initiative for Scientific Excellence (ARISE) funding opportunity from the African Academy of Sciences, one question dominated my thinking: Can agricultural waste be transformed into high-value nanotechnology for health and environmental applications?

This question became the foundation of a multi-year research program and ultimately enabled the establishment of the first dedicated nanotechnology research laboratory in the Democratic Republic of the Congo. Between 2022 and 2026, my students and I explored banana plant waste as a platform for green nanotechnology. What emerged went far beyond nanoparticle synthesis. I will return to my work on silver nanoparticles when the fight against antimicrobial resistance becomes central, but for now I want to briefly highlight our efforts in the transformation and valorization of banana waste through green nanotechnology during the implementation of the ARISE program.

One graduate student, Jonathan M. Mukanya, systematically fractionated banana extracts and identified–to our knowledge, for the first time–the specific phytochemicals responsible for silver ion reduction. This work moved green nanotechnology from empirical practice toward molecular understanding.

Another student, Jimmy K. Kabeya, demonstrated that phytochemical coronas extracted from banana-derived silver nanoparticles exhibited stronger antibacterial activity than either the plant extracts or the nanoparticles alone–revealing a synergistic bio-nano effect. Lamine B. Mona showed that these same biofunctionalized nanoparticles significantly enhanced wound healing, while Joyce K. Mbuyi identified enzymatic activity–such as xylanase-like behavior–in proteins associated with metal nanoparticles, opening new directions in bio-catalytic nanomaterials.

Together, these findings suggested that green nanotechnology is not only a route to affordable materials, but also a powerful discovery engine for biologically active compounds.

Beyond Medicine: One Health in Practice

Green nanotechnology aligns naturally with the One Health framework, which recognizes the interdependence of human, animal, and environmental health. In our work, banana plant waste yielded antimicrobial agents, wound-healing materials, radiation-sensing silica nanoparticles, and low-cost encapsulation systems based on plant-derived saponins.

Each application addresses a different sector, yet all originates from the same principle: waste transformed into value.

Invisible Fixes Made Visible

Green nanotechnology embodies the central message of this book. It transforms what is invisible–discarded materials, neglected ecosystems, underutilized knowledge–into tangible solutions with global relevance. It demonstrates that innovation does not always require more complexity, more energy, or greater expense. Sometimes, it requires seeing value where others see nothing.

By converting agricultural waste into advanced nanomaterials, green nanotechnology offers a pathway not only to affordable healthcare, but also to scientific sovereignty, environmental stewardship, and economic empowerment. This is how nanotechnology can help heal the world–not by replacing nature, but by working with it.

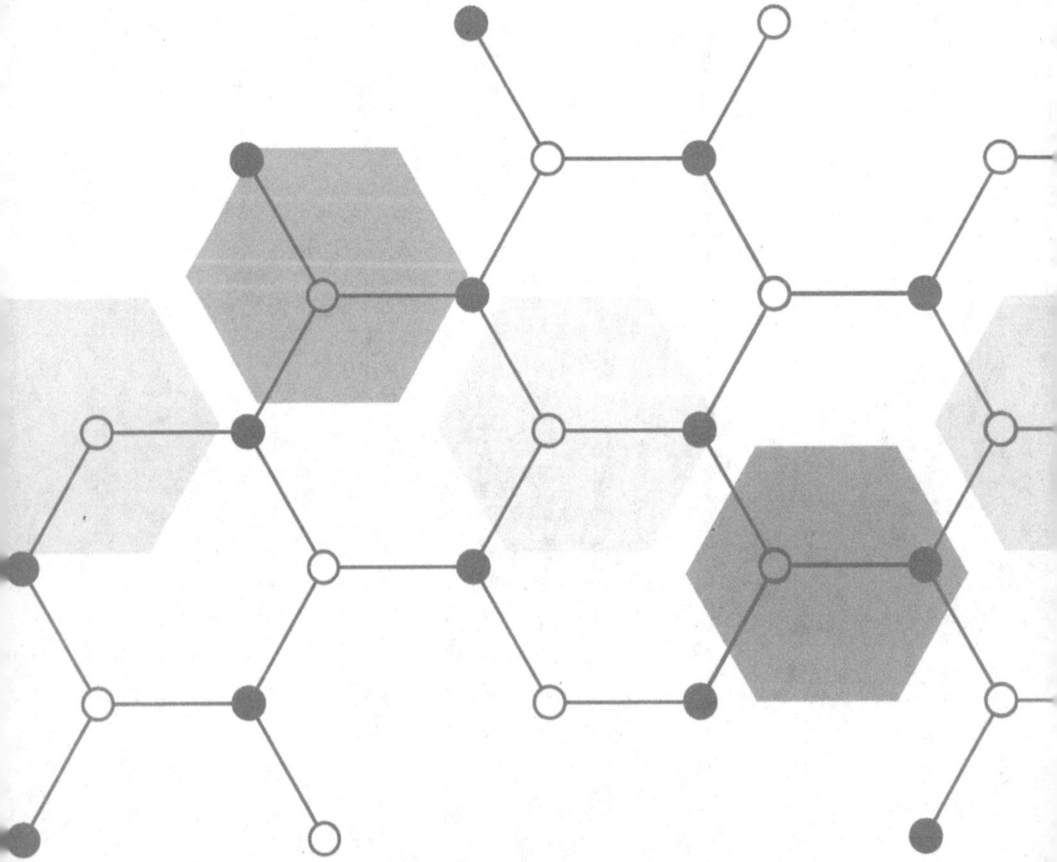

PART III

FROM DISCOVERY TO FIXES

Author's disclaimers note on scientific sources: The content of this part of the book is conceptually drawn on my previously published peer-reviewed research. These sections do not reproduce or summarize any article verbatim, nor do they include figures, tables, or proprietary content from academic publications. Instead, the underlying ideas, insights, and implications of that work have been reinterpreted and rewritten in original language for a general audience, with the goal of public education and scientific accessibility. All interpretations presented here are intended to explain scientific concepts at a high level and should not be construed as clinical guidance or substitutes for primary scientific literature.

CHAPTER 8

GREEN SILVER, LIVING CHEMISTRY:

When Plants Don't Just "Make Nanoparticles," They Program Them

Introduction

Antimicrobial resistance does not announce itself with sirens. It advances quietly—one failed prescription at a time, one wound that no longer heals, one infection that refuses to respond to antibiotics' pressure. For decades, antibiotics carried modern medicine forward, making surgery safer, childbirth survivable, and cancer treatment possible. Today, that foundation is eroding, not because science stopped working, but because the systems that sustained it have reached their limits.

This chapter explores antimicrobial resistance not as a purely biological problem, but as a failure of strategy–one rooted in reductionist thinking, economic imbalance, and global inequity. It then introduces an alternative path: green nanotechnology, where plants and agricultural waste are not passive raw materials, but active chemical architects. Through the lens of silver nanoparticles synthesized from medicinal plants and banana waste, we see how nature can program antimicrobial systems that act through multiple mechanisms at once–confounding resistance rather than provoking it.

These biohybrid materials do not promise a miracle cure. Instead, they reveal a quieter truth: some of the most powerful solutions to modern crises may already exist in overlooked places, waiting to be translated rather than invented.

Antimicrobial Resistance: The Slow Collapse of a Medical Foundation

For much of the twentieth century, antibiotics stood among humanity's greatest achievements. Diseases that once decimated communities–pneumonia, tuberculosis, sepsis, postsurgical infections–became manageable, sometimes routine. Modern surgery, chemotherapy, intensive care, and even safe childbirth quietly depended on the assumption that infections could be controlled. That assumption is no longer secure.

Antimicrobial resistance (AMR) has emerged as one of the defining health crises of our time–not as a sudden catastrophe, but as a gradual erosion of trust in medicine's most basic tools. Bacteria, fungi, parasites, and viruses are evolving faster than our ability to suppress them. Treatments fail where they once succeeded. Infections persist longer, recur more often, and require increasingly toxic or complex interventions. In some cases, no effective therapy remains.

AMR is not a theoretical future. It is already reshaping clinical practice across the world, with disproportionate impact on low- and middle-income countries where diagnostic capacity, alternative therapies, and intensive care resources are limited. The tragedy is not only biological. It is systemic, economic, and deeply inequitable.

How Resistance Emerges: Evolution under Human Pressure

At its core, resistance is not an aberration–it is evolution. Microorganisms replicate rapidly and generate genetic diversity at a pace unmatched by higher organisms. When an antimicrobial agent is introduced, susceptible organisms die, while those with survival advantages persist and multiply. What accelerates this natural process is human behavior.

Antibiotics are frequently misused: prescribed unnecessarily, taken incorrectly, stopped prematurely, or consumed without supervision. In agriculture, antimicrobials are used at a massive scale for growth promotion and prophylaxis. In hospitals, overcrowding, limited diagnostics, and inadequate infection control intensify selective pressure.

Microbes respond with remarkable ingenuity. They:

- Alter drug targets to prevent binding
- Produce enzymes that inactivate antibiotics
- Reduce drug uptake through membrane modifications
- Actively expel drugs via efflux pumps
- Organize into biofilms that shield entire communities

Perhaps most troubling, resistance genes move laterally. Bacteria exchange genetic material across species and environments, meaning a resistance trait that arises in one hospital, farm, or waterway can appear continents away.

AMR is therefore not confined to clinics. It is embedded in ecosystems.

The Antibiotic Innovation Gap

If resistance were balanced by a steady supply of new antibiotics, the crisis might be containable. Instead, antibiotic discovery has slowed dramatically. The reasons are structural. Antibiotics are short-course treatments, tightly regulated by stewardship programs. They are less profitable than chronic disease drugs. As a result, many pharmaceutical companies have withdrawn from antimicrobial research altogether. Startups attempt to fill the gap but often fail before reaching clinical translation, constrained by regulatory complexity, funding limitations, and long development timelines.

The result is a widening gap between microbial evolution and therapeutic innovation.

AMR as a Systems Failure

AMR is often framed as a microbiological problem. In reality, it is a systems failure spanning health, agriculture, environment, economics, and governance. Resistant infections prolong hospital stays, increase mortality, and inflate healthcare costs. They undermine routine medical procedures. Environmental contamination with antimicrobial residues creates reservoirs of resistance. Climate change further exacerbates the problem by reshaping ecosystems and stressing already fragile health systems.

Most importantly, AMR exposes global inequities. While wealthier countries may still access second- and third-line therapies, many regions face infections with no affordable alternatives. Patients die not because science lacks knowledge, but because solutions are inaccessible.

This is the backdrop against which new antimicrobial strategies must be judged–not by novelty alone, but by their capacity to be effective, scalable, and equitable.

Why Conventional Antimicrobial Logic Is Reaching Its Limits

Traditional drug development focuses on single molecular targets. This strategy has yielded remarkable successes, but it also creates vulnerability. Once resistance emerges, efficacy collapses. Escalating doses increases toxicity. Combination therapies increase cost and complexity. Biofilms further complicate treatment. Many pathogens exist not as isolated cells, but as structured communities encased in protective matrices that limit drug penetration and alter metabolism.

The challenge is no longer to kill microbes more aggressively, but to outmaneuver their adaptability.

Why AMR Demands New Ways of Thinking

AMR forces a shift in perspective. Effective solutions must:

- Act through multiple mechanisms simultaneously
- Reduce selective pressure for resistance

- Remain effective against biofilms and persistent cells
- Be affordable and locally adaptable
- Integrate human, animal, and environmental health

This reframes the question: *What counts as an antimicrobial?*

Rather than single molecules, emerging strategies explore complex systems–materials that interact with microbes through physical, chemical, and biological pathways at once.

This is where green nanotechnology enters the narrative.

Green Nanotechnology: When Biology Becomes the Chemist

Green synthesis replaces harsh chemical reagents with biological systems–plants, microbes, enzymes–that act simultaneously as reducing agents and stabilizers.

Plant extracts contain rich mixtures of secondary metabolites–polyphenols, flavonoids, alkaloids, proteins, sugars–capable of reducing metal ions into nanoparticles. But reduction is only the beginning.

Unlike synthetic stabilizers, biological molecules form dynamic organic layers around nanoparticles. These layers influence size, surface charge, aggregation, ion release, and biological interactions.

Plants do not merely produce nanoparticles. They program them.

Silver Revisited: An Old Antimicrobial with New Meaning

Silver has been used for its antimicrobial properties since antiquity. Long before bacteria were understood, silver vessels were used to preserve water, silver coins placed in milk to slow spoilage, and silver compounds applied to wounds.

Yet bulk silver is relatively inert. Its antimicrobial activity arises primarily from silver ions (Ag^{+})–reactive species capable of interacting with biological molecules; silver ions interact with proteins, DNA, and membranes. Traditional silver salts, such as silver nitrate, release ions rapidly and indiscriminately, often causing toxicity alongside antimicrobial effects.

Nanotechnology changed this equation.

At the nanoscale, silver behaves differently. Reducing silver to particles measured in billionths of a meter dramatically increases its surface area, alters its reactivity, and enables controlled interactions with biological systems. Silver nanoparticles (AgNPs) are not merely smaller silver–they are functionally distinct materials.

But not all silver nanoparticles are created equal.

The Problem with Conventional Silver Nanoparticles

Most industrial silver nanoparticles are produced using chemical reducing agents, stabilizers, and organic solvents. While effective at generating particles of defined size and shape, these methods introduce several problems:

- Toxic reagents that require extensive purification
- Environmental burden from chemical waste
- Limited biocompatibility due to synthetic capping agents
- High production costs, restricting accessibility

In many cases, the antimicrobial activity observed is inseparable from cytotoxicity. The nanoparticles kill bacteria–but they may also damage mammalian cells, disrupt ecosystems, or accumulate undesirably in tissues.

For a technology to address AMR meaningfully, it must not only be effective–it must be safe, affordable, scalable, and context-appropriate.

This is where green synthesis fundamentally changes the narrative.

Plant-Derived Silver Nanoparticles: Evidence from Congolese Medicinal Plants

In Kambale, Nkanga, et al. (2020), silver nanoparticles (AgNPs) were synthesized using aqueous extracts from three Congolese medicinal plants: *Brillantaisia patula, Crossopteryx febrifuga*, and *Senna siamea*. The synthesis relied on simple, accessible procedures, aligning with principles of green chemistry. The resulting nanoparticles were spherical and well-dispersed, with average sizes between approximately 45 and 110 nm, as observed by transmission electron microscopy. Characterization using various analytical methods (UV-Vis spectroscopy, FTIR, EDX, and XRD) confirmed the formation of AgNPs and the presence of organic components associated with plant-derived capping agents.

Biologically, these nanoparticles exhibited stronger antibacterial activity than both the corresponding crude plant extracts and silver nitrate against tested bacterial strains. This illustrates a key translational insight: nanoparticle formation can concentrate and reorganize biological activity, transforming weak or diffuse antimicrobial effects into more potent, testable systems.

Importantly, the study of structural neatness (which is scientifically called "crystallinity") suggested that these materials were not pure metallic silver alone but biohybrid systems, incorporating organic matter associated with the plant source. This variability–often dismissed as noise–becomes meaningful when understood and controlled.

From Nanoparticles to Systems: Insights from Banana Plant Waste

The 2025 study by Kabeya et al. extended this logic using banana plant waste (*Musa sapientum and Musa paradisiaca*)–materials typically discarded despite their rich phytochemical content.

Here, the experimental design went further. The study explicitly separated and evaluated: (1) crude plant extracts, (2) biosynthesized AgNPs, (3) AgNPs after removal of organic capping layers, and (4) the isolated organic capping agents themselves. This approach reframed the nanoparticle not as a single entity, but as a two-component system: a metallic core and an organic coat.

Nanoparticle sizes varied depending on measurement technique–approximately 8–40 nm by microscopic visualization (using transmission electron microscopy, "TEM") and 30–140 nm by size measurement in liquid state (using dynamic light scattering, "DLS")–highlighting the importance of multidimensional characterization.

Most strikingly, antibacterial testing revealed that the extracted capping agents from *Musa sapientum* AgNPs showed the highest antibacterial activity, surpassing crude extracts, intact AgNPs, and uncapped nanoparticles.

This finding shifts the narrative decisively. The organic layer is not passive. It can be the dominant antimicrobial contributor.

What This Means for Antimicrobial Strategy

These results suggest that green synthesis does more than produce silver nanoparticles. It acts as a biochemical selection process, concentrating and stabilizing specific phytochemicals at the nanoparticle interface.

The antimicrobial effect becomes:

- Multifactorial rather than single-target
- Physicochemical as well as biochemical
- Less predictable for microbes to resist

Rather than escalating force, such systems confuse microbial adaptation.

Implications for Biofilms and Persistent Infections

Green silver nanoparticles and their capping agents show promise against biofilms–structures that protect bacteria from conventional antibiotics. By disrupting membranes, altering surface interactions, and introducing multiple stress pathways, these systems address one of AMR's most stubborn challenges.

Safety, Equity, and Translational Justice

A recurring criticism of silver-based antimicrobials is toxicity. This concern is legitimate. What distinguishes green silver nanoparticles is not just efficacy, but biological moderation. The phytochemical corona acts as a buffer, slowing silver ion release and reducing nonspecific interactions with host tissues. Green-synthesized AgNPs demonstrate lower cytotoxicity than chemically synthesized counterparts, partly due to moderated silver ion release and biocompatible organic coronas.

Equally important is how they can be produced:

- From locally available plant materials
- Using low-energy, low-cost methods
- Without sophisticated infrastructure

This is essential if silver-based nanomaterials are to move beyond laboratory demonstrations into real-world use. Green nanotechnology therefore opens pathways for decentralized antimicrobial innovation, especially in regions bearing the highest AMR burden.

From Green Chemistry to Invisible Fixes

Green silver nanoparticles exemplify invisible fixes:

- Invisible in scale
- Invisible in origin
- Invisible until failure forces attention

They do not promise a final victory over microbes. But they offer a way to engage biology with humility rather than domination, complexity rather than reductionism.

If antimicrobial resistance is the slow-moving pandemic of our time, then such approaches are not optional innovations. They are necessary adaptations. Thus, the global search for new antibacterial strategies cannot remain confined to a single geography or a single type of laboratory. What makes the antimicrobial crisis especially unfair is that the highest burden of infectious disease often falls on settings that have the least access to expensive drug discovery pipelines. Yet nature–especially medicinal plants used for generations–has always been a distributed library of chemical innovation. The scientific question is no longer whether nature contains antimicrobials; it is whether modern science can translate that chemical wisdom into forms that are more potent, more reliable, and easier to deploy.

CHAPTER 9

TEACHING MEDICINES WHEN TO ACT:

Nanotechnology, Tuberculosis, and the Ethics of Arrival

Introduction

Tuberculosis does not defeat medicine by strength alone. It survives by hiding.

After entering the body, the bacterium responsible for tuberculosis takes refuge inside immune cells that are meant to destroy it. Once inside, it becomes shielded from many antibiotics, forcing patients to endure long treatments with high drug doses that strain the liver, damage other organs, and often fail before the infection is fully cleared.

This chapter explores a simple but powerful idea: Instead of flooding the entire body with medicine, what if we could teach drugs exactly where to go?

Liposomes–tiny particles, with bubble-like structures, made from the same materials as human cell membranes–offer a way to do precisely that. By packaging tuberculosis drugs inside these nanoscale carriers, scientists can guide treatment directly into the immune cells where the bacteria hide. The result is not just improved drug delivery, but a fundamental shift in how tuberculosis therapy can work: more precision, less toxicity, and better chances of success.

By following the journey of liposome-based tuberculosis treatments from laboratory design to biological action, this chapter reveals how nanotechnology transforms old medicines into smarter tools–and how teaching drugs "where to go" can save lives where conventional approaches fall short.

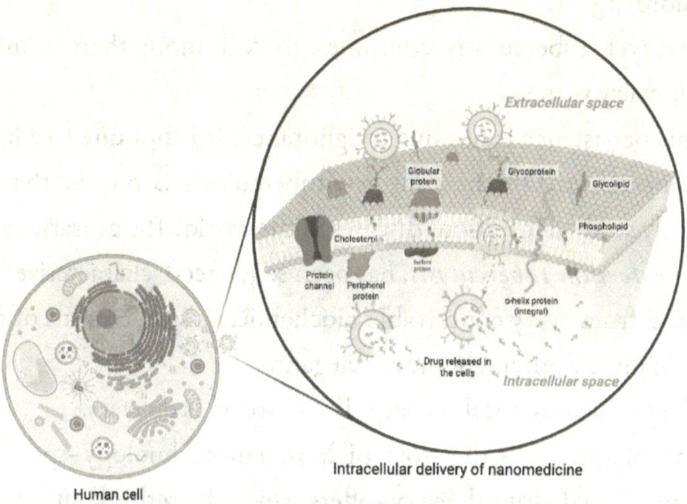

This figure shows how nanomedicine–tiny, engineered particles carrying drugs–travels from outside a cell into its interior. The zoomed-in section focuses on the cell membrane, a thin protective layer made of fats and proteins that carefully controls what can enter or leave the cell. The nanoparticles are designed to recognize and attach to specific parts of this membrane, much like a key fitting into a lock. This allows the cell to take them in, either by wrapping around them

or letting them cross the membrane. Once inside, the particles release their medicine directly, where it is needed. (Courtesy of Mr. Pathy B. Lokole, who prepared this figure using BioRender).

A Disease That Refuses to Stay in the Past

Tuberculosis is often spoken about as a disease of history. In textbooks, it belongs to the nineteenth century. In political speeches, it is framed as a problem already solved by antibiotics. In public consciousness, it is overshadowed by flashier global threats–pandemics that spread faster, cancers that sound more modern, technologies that feel more urgent.

And yet tuberculosis continues to kill more than a million people every year.

This persistence is not due to ignorance. It is not due to a lack of scientific effort. On the contrary, tuberculosis is one of the most intensely studied infectious diseases in the world. The causative agent, *Mycobacterium tuberculosis*, has been sequenced, characterized, and attacked from every conceivable biochemical angle. Dozens of drugs exist. Many are potent. Some are elegant.

The problem is not discovery. The problem is delivery.

This chapter tells the story of how nanotechnology–specifically liposome-based drug delivery–offers a way to rethink tuberculosis treatment by addressing not only what drug is used, but *where, when*, and *how* it acts. More importantly, it shows why this scientific insight cannot be separated from questions of equity, access, and justice. A drug that works in principle but fails to reach the patient who needs it is not a solution. It is an unfinished promise.

Where Tuberculosis Actually Lives

To understand why conventional tuberculosis therapy struggles, one must first understand where the disease hides.

Unlike many bacterial infections that float freely in blood or tissue, *Mycobacterium tuberculosis* is an intracellular pathogen. It survives by invading macrophages–immune cells whose job is to destroy microbes. Once inside, the bacterium manipulates its environment, preventing the macrophage from completing its killing process. The result is a biological paradox: The very cells meant to protect the body become safe houses for infection.

This intracellular lifestyle has profound implications for therapy.

Most antibiotics, including first-line tuberculosis drugs such as isoniazid, circulate freely in the bloodstream. They distribute widely across organs, exposing healthy tissues to their effects, while only a fraction reaches infected macrophages. To compensate, high doses and prolonged treatment regimens are used. The price of this strategy is paid by patients.

The Human Cost of Conventional Therapy

Standard tuberculosis treatment requires daily medication for six months or longer. In some cases–drug-resistant TB–treatment extends to years. Side effects are common: liver toxicity, neuropathy, gastrointestinal distress, fatigue. Patients must continue taking medication long after symptoms improve, often while feeling progressively worse from the treatment itself.

For patients living in resource-limited settings, these burdens are magnified: clinics may be far from home; drug supply may be inconsistent; monitoring for toxicity may be minimal or nonexistent; social stigma may discourage adherence.

Nonadherence is often framed as patient failure. In reality, it is frequently a systemic failure. When therapy demands perfection from people living in imperfect conditions, it is the therapy–not the patient–that is unrealistic.

This is where nanotechnology reframes the problem.

Liposomes: Nanoparticles Speaking the Language of Cells

Liposomes are tiny spherical structures composed of lipid bilayers–the same basic material as cell membranes. Their similarity to biological membranes is not cosmetic; it is functional. Cells recognize liposomes as familiar. Macrophages, in particular, are predisposed to engulf lipid-based particles as part of their natural scavenging behavior.

This biological preference transforms liposomes into delivery vehicles with a built-in targeting mechanism. Without the need for complex ligands or external guidance systems, liposomes naturally accumulate in macrophages–the very cells that harbor tuberculosis bacteria. In this sense, liposomes do not force entry. They are invited.

But targeting alone is not enough. Once inside the macrophage, the drug must remain inactive until it reaches the precise intracellular compartment where the bacteria reside. Premature release wastes the opportunity. Delayed release risks therapeutic failure.

The solution lies not in electronics or external control, but in chemistry.

pH: An Invisible Signal with Powerful Consequences

In simple terms, pH can be defined as a measurement scale that tells how acidic or basic (alkaline) a water-based liquid is; it can be considered a sourness or caustic meter for liquids, ranging from 0 to 14.

Inside the human body, pH varies subtly but meaningfully. Blood and extracellular fluids are close to neutral (pH around 7). Inside cells, especially within compartments of macrophages, like endosomes and lysosomes, the environment becomes progressively acidic (pH < 7).

Macrophages exploit this acidity to kill pathogens. Ironically, *M. tuberculosis* has evolved mechanisms to survive in these acidic compartments. But the acidity itself remains a reliable signal–a kind of molecular "address label."

My research sought to use this pH difference as a trigger.

The idea was simple in principle: Design a drug–carrier system that remains stable at neutral pH but disassembles or releases its cargo when exposed to acidic conditions. This would ensure that the drug in liposomes remains protected during circulation and is only released after the liposome has been taken up by the macrophage.

Engineering Intelligence into Molecules

Isoniazid is one of the cornerstones of tuberculosis therapy. It is highly effective, inexpensive, and widely available. Yet it also illustrates the limitations of conventional drug design. Isoniazid diffuses rapidly throughout the body, exposing the liver to toxicity while failing to accumulate selectively inside infected macrophages.

Rather than inventing an entirely new drug, I chose a more strategic approach: re-engineering an existing one.

In my work, isoniazid was chemically modified to introduce pH sensitivity. The modification relied on hydrazone linkages, bonds that are stable at neutral pH but break under acidic conditions. This molecular modification of isoniazid led to formation of hydrazone conjugates. In these new constructs, the hydrazone bonds act like molecular switches, responding to pH changes with remarkable sensitivity.

As explained earlier, this matters because the intracellular compartments of macrophages (phagosomes, lysosomes, and phagolysosome), where *Mycobacterium tuberculosis* resides, are acidic. The macrophage's phagolysosome is not just a container; it is a chemically distinct environment.

By linking isoniazid to a hydrophobic anchor via a pH-sensitive bond and incorporating this construct into a liposome, I created a system with layered intelligence:

- Targeting intelligence through macrophage uptake
- Temporal intelligence through delayed release
- Environmental intelligence through pH-triggered activation

This modification did not alter the drug's therapeutic target. Instead, it altered the timing and location of drug activation. In essence, the drug was taught to wait. It remains largely silent during circulation (at neutral pH 7). It speaks only when it reaches the right place (inside macrophages, pH < 7).

For the nonscientific reader, the message is simple but powerful: Tuberculosis persists not because we lack drugs, but because we lack delivery wisdom. Nanotechnology offers that wisdom–not by replacing medicine, but by teaching it how to behave.

Controlled Release as Respect for the Body

In lay discussions, controlled release is often simplified to "slow release." This framing misses the point. Controlled release is not about slowing time. It is about respecting context.

A well-designed controlled-release system acknowledges that the body is not uniform. It recognizes that tissues differ, cells differ, and intracellular environments differ. Rather than overwhelming the body with excess drug, it delivers precisely what is needed, where it is needed. For patients, this translates into lower systemic toxicity, reduced dosing frequency, improved tolerability, and greater dignity during treatment.

These are not minor conveniences. They are determinants of whether therapy succeeds in the real world.

Patient-Centered Implications: Beyond Pharmacology

From a patient's perspective, the most revolutionary aspect of nanotechnology-enabled tuberculosis therapy is not its novelty–it is its humaneness.

Imagine a treatment regimen where:

- Medication is taken less frequently
- Side effects are significantly reduced

- Treatment aligns with daily life rather than dominating it
- The burden of adherence is shared by the system, not placed entirely on the patient

Such a regimen changes the psychological experience of illness. It transforms treatment from punishment into partnership.

This matters especially in diseases like tuberculosis, where stigma, fatigue, and economic precarity intersect. A therapy that reduces clinic visits and physical suffering does more than cure infection–it preserves social participation.

Why This Science Rarely Reaches the People Who Need It Most

At this point, an uncomfortable question arises: If nanotechnology-based solutions are scientifically sound, why are they not already standard of care? The answer lies in translation.

Nanomedicine does not fail because it lacks brilliance. It fails because of the complexity of translation, requiring infrastructure, regulatory expertise, manufacturing capacity, and sustained investment. These elements are unevenly distributed across the globe. As a result, many of the most elegant solutions remain confined to laboratories in high-income countries. Patients in low- and middle-income regions–where tuberculosis is most prevalent–are left waiting.

This is not a failure of science. It is a failure of circulation.

This chapter cannot be separated from the broader thesis of this book: that science advances most powerfully when knowledge moves, returns, and adapts. This chapter belongs in this book because it embodies its central thesis: The smallest interventions, when intelligently designed and equitably deployed, can produce the greatest healing. The liposomes described here are invisible to the eye. But their implications–for patients, health systems, and global equity–are anything but.

Translational Justice: When Discovery Is Not Enough

Translational justice demands that scientific advances do not remain trapped in privileged spaces. It asks difficult questions:

- Who benefits from innovation?
- Who decides which problems are worth solving?
- Who controls the path from bench to bedside?

Nanotechnology offers tools. Brain circulation offers pathways. But justice requires intention. A tuberculosis nanomedicine that never leaves the laboratory is not neutral. It is an opportunity withheld.

This chapter, therefore, is not merely a technical explanation. It is an ethical argument: that the success of science must be measured not by publications, but by arrival–by whether solutions reach the people whose lives depend on them.

A Quiet Revolution with Loud Implications

What began as an exercise in chemical engineering becomes, upon reflection, a blueprint for a different kind of science. One that:

- Designs with patients in mind
- Builds intelligence into materials
- Respects biological complexity
- Refuses to accept geographic inequality as inevitable

In teaching medicines when to act, nanotechnology teaches scientists when to listen–to biology, to patients, and to the world beyond the laboratory.

CHAPTER 10

WHEN A VIRUS BECOMES A MEDICINE:

Reengineering Nature to Fight Cancer

Introduction

Viruses have long been feared as agents of disease–symbols of contagion, uncertainty, and loss. Yet beneath this reputation lies a deeper truth: Viruses are among nature's most efficient delivery systems. They are experts in targeting cells, navigating biological barriers, and triggering powerful immune responses. What if these same properties could be redirected–not to cause disease, but to treat it?

This chapter explores how plant viruses, reengineered at the nanoscale, can become allies in the fight against cancer. Being noninfectious to mammals, these viral nanoparticles act as precision tools–carrying therapeutic cargo, activating the immune system, and reshaping the tumor microenvironment. By learning from nature rather than fighting it, scientists are transforming ancient biological machines into modern medicines.

Through the example of tobacco mosaic virus–based nanoplatforms, this chapter reveals how viral architecture can be repurposed to deliver immunotherapies and light-activated treatments that selectively destroy cancer cells. In doing so, it challenges a fundamental assumption of medicine: that danger must always be eliminated, rather than redesigned. Sometimes, the most powerful cures come from what we once feared the most.

A Virus, Reconsidered

For most of human history, viruses have been symbols of fear. They are invisible enemies–agents of epidemics, devastation, and loss. From influenza pandemics to HIV, Ebola, and COVID-19, viruses have shaped our collective trauma and exposed the fragility of global health systems.

Yet science, at its best, teaches us a difficult but powerful lesson: Nothing in biology is inherently good or evil. Everything depends on context, intention, and design.

The tobacco mosaic virus (TMV) is one of the oldest viruses ever identified. First described at the end of the nineteenth century, it devastated tobacco crops and became a textbook example of plant pathology. For decades, its role was purely destructive–an agricultural threat to be contained.

But TMV also became one of the most studied biological structures in history. Its rigid, rod-shaped architecture fascinated scientists long before nanotechnology existed as a discipline. When the language of nanoscale engineering finally emerged, TMV was waiting.

This chapter tells the story of how TMV–once a symbol of biological harm–was transformed into a precision nanomedical tool. More importantly, it explains why this transformation matters not only scientifically, but ethically, socially, and globally. Because when a virus can be turned into a medicine, it forces us to ask a deeper question: What else have we failed to reimagine?

Why Delivery Matters More Than Drugs

Most people believe medicine fails because we lack powerful drugs. In reality, medicine often fails because we cannot deliver those drugs correctly.

Chemotherapy agents can kill cancer cells–but they kill healthy cells too. Immunotherapies can awaken the immune system–but sometimes unleash dangerous inflammation. Light-based therapies can destroy tumors–but only if energy reaches the right depth, with surgical precision.

The real challenge is not discovering molecules. It is controlling *where, when,* and *how* those molecules act.

This is the core promise of nanomedicine: not stronger drugs, but smarter behavior.

Instead of flooding the body with toxic chemicals, nanomedicine aims to guide therapeutic agents to specific tissues, protect them in circulation, activate them only when needed, and remove them safely afterward.

TMV offers a rare opportunity to do all of this in one platform.

TMV as a Nanoscale Scaffold

TMV is not spherical like many synthetic nanoparticles. It is a rigid nanorod–approximately 300 nm long and 18 nm wide–assembled from more than two thousand identical protein subunits. This uniformity is not cosmetic. It is functional.

Every TMV particle is nearly identical to the next, produced by nature with atomic precision. In a field where batch-to-batch variability undermines reproducibility and regulatory approval, this is revolutionary.

Equally important: TMV does not infect humans. It cannot replicate in mammalian cells. It cannot integrate into human DNA. It is biologically inert in the human body–unless we choose to give it a function. This makes TMV an ideal *blank canvas.*

Turning TMV into a Diagnostic Tool

One of the earliest transformations involved turning TMV into a medical imaging agent.

Magnetic resonance imaging (MRI) relies on contrast agents such as gadolinium to visualize tissues. But conventional gadolinium agents circulate quickly, require relatively high doses, and offer limited sensitivity.

By loading gadolinium ions into the inner channel of TMV, we created a nanoscale contrast agent where thousands of imaging atoms travel together, dramatically amplifying signal strength.

This does two things simultaneously:

- Improves diagnostic sensitivity
- Reduces the amount of gadolinium required

For patients, this translates into safer imaging with clearer results–especially important for individuals requiring repeated scans, such as cancer patients under long-term monitoring.

But diagnosis alone is not enough. The next transformation was more radical.

Teaching a Virus to Respond to Light

We coated TMV particles with polydopamine, a biocompatible polymer inspired by the adhesive chemistry of mussels. Polydopamine absorbs near-infrared light and converts it into heat–a property that can be harnessed for photothermal therapy, a treatment approach where a product converts light (often near-infrared) into heat, selectively killing diseased cells in the surrounding area. Photothermal therapy works on a simple principle: Heat kills cancer cells. But uncontrolled heat damages healthy tissue.

TMV-based photothermal systems solve this by localizing heat generation. Only where TMV accumulates–and only when light is applied–does heat emerge.

This gives clinicians something unprecedented:

- **Spatial control** (where treatment occurs)
- **Temporal control** (when treatment occurs)
- **Dose control** (how intense treatment is)

Light becomes a remote switch. The patient does not feel the virus working. They do not feel the nanoparticles circulating. The fix remains invisible–until it acts.

Immunotherapy: Training the Body, Not Replacing It

Cancer does not exist in isolation. It survives by hiding from the immune system.

One of the most powerful insights from immunotherapy is that long-term cancer control depends on immune memory. Destroying a tumor is not enough; the immune system must learn to recognize it.

TMV is naturally immunogenic–not dangerously so, but enough to act as an immune stimulant. When engineered correctly, it becomes a training platform for the immune system.

By combining TMV-based immunotherapy with photothermal tumor destruction, we observed a powerful synergy:

- Tumor cells were destroyed locally.
- Tumor antigens were released.
- The immune system, already activated, responded systemically.

This is how local treatment becomes global protection.

In preclinical melanoma models, this combination not only shrank tumors but reduced recurrence–suggesting immune memory formation.

For patients, this matters more than tumor size. It means time, hope, and durability.

The Body's Response: Protein Coronas and Biological Negotiation

Nanomedicine does not operate in a vacuum. The moment a particle enters the bloodstream, it is interpreted by the body. Proteins attach to its surface, forming a protein corona that determines how the particle is perceived. This process is often unpredictable and can lead to rapid clearance or immune reactions.

TMV provided a surprising advantage: The polydopamine coating altered the protein corona composition, reducing aggressive immune recognition while preserving circulation time. The virus did not hide–it adapted.

This study taught us a critical lesson: Successful nanomedicine does not fight biology; it collaborates with it.

A Virus as a Symbol of Justice

There is something profoundly poetic about using a virus–a symbol of uncontrolled spread–as a tool for controlled healing. TMV does not replicate uncontrollably. It does not invade blindly. It moves with purpose. So should science.

TMV teaches us a metaphor: Just as a therapeutic payload must be carried by a vehicle to reach its target, knowledge must be carried by people to reach society. Invisible fixes are not just nanoscale tools. They are systems of movement–of ideas, people, and responsibility. When scientists migrate permanently, capacity drains. When they circulate–training abroad, returning home, collaborating across borders–capacity multiplies.

The technologies described in this chapter were experimented during my movement in laboratories across the globe, where ideas were exchanged through mobility, expertise accumulated through circulation.

But circulation must be intentional. Since cancer nanomedicine matters for patients globally, including in low- and middle-income countries, it is necessary to not just publish such research, but also build local translational capacity to ensure global impact.

CHAPTER 11

WHEN TIME BECOMES THE MEDICINE:

Hydrogels, Viral Nanoparticles, and the Architecture of Sustained Immunity

Introduction

Most medicines act in moments. They enter the body, reach a peak, and fade. But immunity does not unfold in moments–it is built over time. The immune system learns slowly, remembers selectively, and responds through repeated encounters. When therapies ignore this temporal logic, even the most sophisticated drugs can fall short.

This chapter explores a different design principle: what happens when time itself becomes part of the medicine. By combining viral nanoparticles with biocompatible hydrogels (which are soft, water-filled materials), scientists can shape not only what the immune system sees, but when and how often it sees it. The result is not a stronger stimulus, but a wiser one–an immune response that matures, deepens, and persists. Modern medicine has long been obsessed with potency.

Stronger drugs. Faster responses. Higher doses. Immediate effects. In emergency medicine, this instinct saves lives. But in immunology–in cancer, vaccination, and chronic disease–speed is not always the answer. Sometimes, duration is more powerful than intensity. Sometimes, how long the immune system is engaged matters more than how hard it is pushed.

Through examples drawn from cancer immunotherapy and vaccine development, including my work on cowpea mosaic virus-based platforms, this chapter reveals how sustained delivery transforms transient exposure into durable protection. In these systems, the medicine does not rush. It stays. And in staying, it teaches the immune system how to defend itself long after the injection is forgotten.

This chapter is about a quiet but transformative idea: that time itself can be engineered as a therapeutic variable. By combining virus-based nanotechnologies with injectable hydrogels, we learned how to slow down immune instruction, allowing the body to learn, adapt, and remember.

The Hidden Weakness of Modern Immunotherapy

Immunotherapy is often described as one of the greatest medical breakthroughs of the twenty-first century. Checkpoint inhibitors, cancer vaccines, and immune-stimulating nanoparticles have reshaped oncology and infectious disease research. Yet beneath these successes lies an uncomfortable truth: Many immune-based therapies fail not because they are weak, but because they disappear too quickly.

The immune system does not respond like a light switch. It responds like a classroom. Cells must encounter a signal repeatedly, in the right context, over time. Dendritic cells need sustained exposure to antigens to mature properly. T cells require prolonged stimulation to differentiate into memory cells. Abrupt, transient exposure often leads to partial activation–or worse, immune tolerance.

In practical terms, this means that many promising immunotherapies demand frequent injections, repeated boosters, or continuous infusions. These requirements introduce three systemic problems:

1. **Patient burden:** frequent hospital visits, injections, and monitoring
2. **Health system strain:** dependence on specialized infrastructure
3. **Global inequity:** therapies become inaccessible outside well-resourced settings

This is where the science quietly breaks down–not at the level of molecules, but at the level of delivery.

Plant Virus Nanoparticles: Nature's Immune Alarm Systems

Among the most elegant immune activators discovered by nanoscience are plant virus nanoparticles. Unlike human viruses, these particles are noninfectious to humans. Yet they retain highly ordered, repetitive protein structures that the immune system instinctively recognizes as foreign. Cowpea mosaic virus (CPMV) is one such platform. It is structurally precise, genetically stable, and naturally immunogenic. When introduced into mammalian tissue, CPMV activates innate immune pathways, recruiting macrophages, dendritic cells, and natural killer cells. These early signals then propagate into adaptive immune responses capable of reshaping tumor microenvironments and enhancing immune surveillance.

In cancer models, CPMV has demonstrated the ability to:

- Convert immunologically "cold" tumors into "hot" ones
- Increase antigen presentation
- Promote cytotoxic T cell infiltration

Yet CPMV also illustrates a recurring translational problem: rapid clearance. When administered in soluble form, CPMV is quickly eliminated from the body. Its immune signal is sharp but fleeting.

This creates a paradox. The nanoparticle is biologically powerful–but pharmacologically fragile.

A Shift in Perspective: Engineering the Environment of the Virus

Many approaches attempt to solve rapid clearance by modifying the nanoparticle itself—PEGylation, surface shielding, chemical alteration. While effective in some contexts, these strategies risk dampening immunogenicity or introducing toxicity. We chose a different path.

Instead of asking, *How do we make CPMV last longer in the bloodstream?*, we asked, *What if we changed the way the immune system encounters CPMV altogether?*

This reframing led us to injectable hydrogels–soft, water-rich polymer networks capable of forming localized depots within the body. Hydrogels do not chase circulation. They stay put, releasing their cargo gradually into surrounding tissue.

In other words, hydrogels allow therapy to linger, not circulate.

Hydrogels as Time-Architects

The hydrogels used in our work were based on chitosan and glycerophosphate, materials already known for their biocompatibility and clinical safety. At room temperature, the formulation behaves like a liquid. Upon injection into the body, physiological temperature triggers gelation, forming a semi-solid matrix at the injection site.

This transformation is deceptively simple–but biologically profound.

Once CPMV is embedded within the hydrogel, it is released slowly over weeks. Importantly:

- The virus particles remain structurally intact.
- Their immunogenic surface features are preserved.
- The immune system encounters them continuously, rather than transiently.

The immune system is no longer shocked–it is educated.

Preclinical Evidence: One Injection, Sustained Immunity

In murine colon cancer models, this sustained delivery strategy produced striking results. A single hydrogel-based administration of CPMV generated immune activation comparable to–or exceeding–

that of multiple injections of soluble CPMV. Tumor growth was significantly suppressed. Immune cell infiltration was prolonged. Local inflammation was controlled rather than chaotic. Most importantly, the immune response matured rather than spiked. From a scientific perspective, this confirmed that kinetics matter. From a translational perspective, it revealed something even more important: Fewer injections can produce stronger outcomes.

From Treatment to Prevention: Teaching the Immune System to Remember

One of the most compelling aspects of sustained CPMV delivery was its impact on immune memory. When immune stimulation is extended, the body does not merely react–it learns.

This shifts immunotherapy from a reactive intervention to a preventive strategy. Rather than chasing tumors after they arise, the immune system becomes better prepared to recognize malignant transformations early.

In this sense, hydrogels transform virus nanoparticles from weapons into teachers.

Implications for Vaccines and Pandemic Readiness

The same principles that enhance cancer immunotherapy apply directly to vaccination.

Vaccines often fail not because they are ineffective, but because:

- Booster schedules are not completed
- Cold-chain logistics break down

- Repeated access to clinics is not feasible

A single-dose vaccine capable of sustaining antigen exposure could dramatically improve global coverage. During the COVID-19 pandemic, the world learned how fragile vaccine distribution systems can be. Technologies that reduce dependence on repeated dosing are not luxuries–they are necessities.

Hydrogel-based vaccine depots offer a way to:

- Simplify immunization schedules
- Reduce healthcare burden
- Improve equity across regions

Why This Matters in Low-Resource Settings

In many parts of the world, access to advanced therapies is not limited by scientific knowledge, but by delivery feasibility. A therapy that requires repeated hospital visits reinforces inequality. A therapy that works from a single administration levels the playing field.

Hydrogels are particularly well suited to this challenge because they:

- Use relatively simple materials
- Do not require complex devices
- Can be locally manufactured with appropriate training

This aligns directly with the philosophy of translational justice that runs throughout this book.

Personal Reflection: When Science Is Too Far Away

My advocacy for translational development is not abstract. It is personal.

When my brother died in 2019 from a liver disease deemed untreatable, what haunted me was not only loss–it was silence. Somewhere, therapies existed. Somewhere, research had advanced. But those solutions had not traveled far enough, fast enough, or fairly enough.

Working on sustained immunotherapy forced me to confront an uncomfortable truth: Science can be brilliant and still fail people if it is not designed with access in mind.

Hydrogels are not just delivery systems. They are reminders that innovation must meet reality.

Policy Implications: Designing for Equity from the Start

This work raises urgent questions for funders, regulators, and policymakers: Why are delivery systems often treated as secondary considerations? Why is scalability rarely rewarded at early research stages? Why do we fund molecules without funding pathways to people? If sustained-delivery platforms were prioritized earlier, fewer therapies would stall between bench and bedside.

Toward the Future of Invisible Fixes

The most powerful fixes are rarely dramatic. They are quiet, structural, and patient.

- Nanoparticles that do not rush
- Hydrogels that hold time
- Immune systems allowed to learn slowly
- Knowledge that travels and returns

Together, these elements redefine what it means for science to heal the world.

Not by force–but by design.

CHAPTER 12

BEYOND THE BENCH:

Building Translational Capacity Where It Matters

Introduction

Innovation is not failing because science lacks ideas. It is failing because ideas too often stop where impact should begin.

Across laboratories worldwide, discoveries emerge at an unprecedented pace. Breakthroughs are published, celebrated, cited–and then quietly archived. They remain elegant demonstrations of what *could* be done, rather than durable solutions that are done. Between discovery and delivery lies a vast, underdeveloped territory: translation.

This chapter argues that scientific excellence alone is no longer sufficient. If research is to serve society meaningfully, it must be accompanied by deliberate investment in translational capacity–the skills, systems, and infrastructures that transform knowledge into usable, scalable, and equitable solutions. Without this bridge, even the most promising innovations risk becoming intellectual artifacts rather than instruments of change.

The imbalance is stark. Millions of scientific papers coexist with persistent therapeutic scarcity, particularly in regions where disease burden is highest. This contrast does not reflect a failure of imagination or a shortage of effort. It reflects a structural gap: Science is trained to generate answers, but not always to deliver them. Documentation falters, scalability is postponed, regulatory realities arrive too late, and end users are consulted only after critical decisions have already been made.

Translation is not a final step–it is a discipline. One that must be learned, practiced, and embedded early. It requires thinking beyond the bench, beyond the experiment, and beyond publication. It demands fluency in quality systems, development pathways, and real-world constraints. Most importantly, it demands people trained to carry ideas forward, not just to discover them.

In this chapter, I explore why strengthening local and regional translational capacity is essential if science is to move beyond academic success and toward societal relevance. Drawing from lived experience across research laboratories, startups, and development partnerships, I argue that equitable impact will remain elusive until we train not only more scientists–but more translators of science.

Invisible fixes do not fail because they are invisible. They fail when no one is equipped to carry them forward.

Every year, laboratories across the world generate discoveries of extraordinary sophistication. New materials, new molecules, new platforms, new therapeutic concepts–many of them elegant, creative, and genuinely promising. Academic publishing reflects this abundance. The volume of scientific literature has grown exponentially, particularly in fields such as nanotechnology. According to StatNano, more than 1.8 million nanotechnology-related publications were produced between 2000 and 2019 alone.

Yet, paradoxically, there is striking disconnect from the clinic that suggests a field that is producing more knowledge but converting less of it into impact. Despite millions of published studies and decades of research investment, fewer than one hundred nanotechnology-based products had reached regulatory approval worldwide by 2020–suggesting only < 0.005% happened to become useful for the patient. This gap is not merely statistical–it is human. It is measured in patients who die from diseases that science already knows how to treat, in communities that never benefit from discoveries made in their name, and in researchers who see their work end at the publication stage rather than at the bedside.

I have lived on both sides of this divide. I have watched people lose their lives to cancers labeled "non-curative," while simultaneously working in laboratories where nanoparticles successfully prevented or eradicated tumors in preclinical models. The contrast is devastating. It forces an uncomfortable but necessary question: If the science exists, why does the solution not reach the people who need it most?

The answer is not a lack of brilliance, nor a lack of initiative. It is the absence of translational capacity.

The Global Bottleneck: When Innovation Stalls

Across the innovation ecosystem, failure is often misattributed. Startups are blamed for poor execution, academics for being "too theoretical," regulators for being "too slow." But these explanations miss the deeper structural issue. Translation–the process of turning discovery into deployable solutions–is a discipline in its own right. And it is one that remains underdeveloped, under-taught, and unevenly distributed across the world.

Health technology investors estimate that 60 to 80 percent of health-tech startups fail before reaching clinical adoption, not because the underlying science is flawed, but because they cannot navigate regulatory, clinical, manufacturing, or reimbursement hurdles. This phenomenon–often referred to as *pilot purgatory*–traps innovations between proof-of-concept and real-world use. They work well enough to publish, but not well enough to scale.

This is not accidental. Most scientists are trained to generate knowledge, not to operationalize it. Yet society increasingly expects research to deliver tangible outcomes: therapies, diagnostics, devices, policies, and systems that improve lives. Without deliberate investment in translational skills, this expectation becomes unfair–and ultimately unfulfilled.

The First Translational Failure: Documentation

The most overlooked barrier to translation is also the most fundamental: how science is recorded.

Good Documentation Practices (GDP) are not glamorous. They do not generate citations. They are rarely taught explicitly in academic curricula. Yet they are the foundation upon which all translation rests. Poor documentation destroys reproducibility, and without reproducibility, there can be no scale, no validation, no regulatory approval.

A landmark survey published by *Nature* in 2016 revealed a sobering reality: More than 70 percent of scientists reported being unable to reproduce the work of others, and over half could not reproduce their own experiments. This is not a failure of integrity–it is a failure of training and systems.

In translational development, every parameter matters. Batch numbers, material grades, environmental conditions, procedural deviations–all of these details become critical once a product moves beyond exploratory research. A protocol that works once in a laboratory may fail catastrophically when repeated under regulatory scrutiny. Translation does not tolerate ambiguity.

Building local translational capacity therefore begins with something deceptively simple: teaching scientists to document as if their work will one day be audited, manufactured, and used in humans. Because if it cannot survive documentation, it cannot survive translation.

Scalability Is Not an Afterthought

The second major translational barrier is scalability–and it is one that must be addressed from the very first experiment.

In many academic settings, research is optimized for feasibility, speed, and cost reduction. Research-grade reagents are used because they are accessible. Equipment is chosen because it is available. Processes are designed to work at milligram scale, under conditions that may never be replicated in manufacturing environments.

But when a discovery moves toward development, everything changes.

I have experienced this transition firsthand. Formulations that performed perfectly with research-grade materials required complete re-optimization when switched to good manufacturing practice (GMP)-grade inputs. I remember my fellow scientists calling me: "Christian, this GMP material has nothing to do with what we used last time. It's giving us a highly viscous fluid that we can't process further as usual." We had to revisit the formulation: Polymer concentrations had to be reduced sixfold. Device designs had to be reengineered to accommodate new physicochemical properties. Equipment that functioned well at bench scale proved incompatible with validated manufacturing workflows.

None of these challenges invalidates the science–but they expose a critical gap between discovery and deployment.

Scalability is not about making something bigger. It is about making it repeatable, controllable, and compliant. This requires early decisions about material selection, process robustness, analytical methods, and equipment qualification. Without local expertise in these areas, promising innovations stall–or are exported elsewhere for development, reinforcing global inequities.

Translation Begins with the End User

Another frequently neglected principle of translation is early and continuous engagement with end users.

Too often, research begins with a technique rather than a need. Scientists ask what can be made, not what must be solved. As a result, solutions are developed in search of problems, rather than the reverse. When translation is attempted later, misalignment becomes apparent: The product does not fit clinical workflows, regulatory pathways, or economic realities.

Unmet need-driven research changes this trajectory. It requires listening–to clinicians, patients, pharmacists, regulators, and communities–before the first experiment is run. It demands humility and openness, but it dramatically increases the likelihood of adoption. In regions where healthcare systems face resource constraints, this principle is even more critical. Solutions must be not only effective but affordable, maintainable, and context-appropriate. Translational capacity therefore includes the ability to ask the right questions early, not just to generate sophisticated answers later.

The Hidden Complexity of Translation

Perhaps the greatest misconception about translational development is that it is simply "applied research." In reality, it is a multidisciplinary orchestration involving quality systems, regulatory strategy, manufacturing science, clinical design, supply chains, and market access.

Most small research groups and startups do not fail because they lack intelligence–they fail because they attempt to do everything themselves.

One of the most effective translational strategies is selective outsourcing. Contract Research Organizations (CROs), Contract Development Organizations (CDOs), and Contract Development and Manufacturing Organizations (CDMOs) exist precisely to bridge capability gaps. When used wisely, they accelerate development, reduce risk, and provide access to specialized expertise that would be impossible to build internally.

However, outsourcing is not a panacea. It must be strategic. Research itself–exploratory, uncertain, iterative–does not always fit well within outsourced models. Development, validation, and scale-up, on the other hand, often do. Knowing what to outsource, when, and to whom is itself a translational skill–one that must be taught.

Rethinking the Translational Ecosystem

My experience working with CROs, CDOs, and CDMOs has revealed both their value and their limitations.

Many operate under time-and-materials or cost-plus models, which protect service providers but place disproportionate risk on innovators. Projects may consume significant resources without guaranteeing progress. This misalignment of incentives can quietly drain innovation pipelines.

There is room for evolution. Milestone-based or outcome-linked models, particularly for early-stage innovation, could better align accountability and reward meaningful advancement rather than activity alone.

Similarly, the breadth of services offered by many CDMOs–while impressive–can obscure areas of true excellence. Strategic specificity, transparent referral networks, and earlier engagement with emerging science could transform these organizations from service vendors into ecosystem enablers.

Finally, the translational workforce itself must evolve. Experience is invaluable, but so is future capacity. Training the next generation of translational scientists–people fluent in both science and systems–is essential if innovation is to reach society at scale.

Why Local Translational Capacity Matters

For low- and middle-income regions, the stakes are even higher. When translational capacity is absent locally, discoveries are either abandoned or exported. In both cases, the region loses–not only potential therapies, but agency.

My efforts to establish nanotechnology research in the Democratic Republic of the Congo–through programs like ARISE and the creation of CReNAPN–were never meant to stop at publications. They were intended as a foundation. But research capacity alone is not enough. Without parallel investment in translational training, quality systems, regulatory literacy, and development partnerships, discoveries will continue to stall.

If science is to serve society, it must be able to move.

Training Translators of Science

The future of equitable innovation depends on a new kind of professional: the translator of science.

These are individuals trained beyond the bench–fluent in experimental rigor, documentation, scalability, regulation, and user-centered design. They do not replace scientists or clinicians; they connect them. They ensure that knowledge does not remain trapped in journals but flows toward impact.

Training such professionals must become a global priority. Not only in elite institutions, but in regional universities, research centers, and innovation hubs. Translation must be taught, practiced, and valued as a discipline–not treated as an afterthought.

Only then can science fulfill its promise.

From Discovery to Dignity

Innovation without translation is incomplete. Discovery without delivery is inequitable.

If we truly believe that science exists to improve lives–not just to advance careers–then we must invest in the systems, skills, and structures that allow knowledge to move. Enhancing local and regional translational capacity is not optional. It is the missing link between brilliance and benefit.

This chapter is not a critique of science. It is a call to complete it.

Invisible fixes do not heal the world by being clever. They heal it by being carried–carefully, deliberately, and equitably–from the bench to the people who need them most.

CHAPTER 13

NANOTOXICOLOGY:

Making Sure Invisible Fixes Do Not Become Invisible Harms

Introduction

Nanotechnology has taught us how to act with precision at a scale biology understands. Nanotoxicology teaches us how to act with restraint at a scale society demands.

In medicine, success is often measured by what disappears–a tumor shrinks, an infection clears, a viral load drops. But nanomedicine introduces a deeper question: What remains after the solution works? Nanotoxicology exists to answer that question before harm becomes history.

This chapter is not a pause on innovation. It is a safeguard for it. Because nanoparticles are not merely smaller drugs–they are materials with memory, movement, and consequence. Their size allows them to circulate longer, interact differently with immune systems, and persist in tissues in ways conventional medicines never could. These same advantages that make nanotechnology powerful also make safety nonnegotiable.

Using familiar platforms–liposomes, iron oxide nanoparticles, and biodegradable polymeric systems–this chapter explores how safety is engineered, not assumed. It shows how immune activation can heal or harm, how degradation can liberate or irritate, and how time itself becomes a biological variable that must be designed for rather than ignored.

Nanotoxicology Overview

Nanotechnology promises precision. It promises elegance. It promises solutions that are smaller, smarter, and more selective than anything medicine has known before. Throughout this book, we have seen how nanoparticles can teach medicines where to go, how they can transform viruses into vaccines, waste into weapons against antimicrobial resistance, and time itself into a therapeutic tool. But there is a question that must always follow promise:

What happens after the miracle works?

Nanotoxicology exists to ask that question honestly.

This chapter is not written to slow innovation. It is written to *protect* it. Because history has taught us a hard lesson: Technologies that advance faster than trust eventually collapse under their own success. Nanotoxicology is the discipline that ensures nanotechnology does not repeat the mistakes of past medical revolutions–where solutions arrived before society understood their long-term consequences.

Why Nanotoxicology Is Not an Obstacle–But a Bridge

The word *toxicology* often evokes fear. Poison. Danger. Harm. But modern toxicology–and especially nanotoxicology–is not about alarmism. It is about prediction, prevention, and design. Nanotoxicology allows scientists to ask not only, "Does this kill cells?" but also, "Which cells? Under what conditions? At what dose? For how long? And what happens next?"

These questions matter profoundly when we are designing materials meant to circulate through blood, lodge in tissues, interact with immune cells, or persist long enough to deliver sustained benefit. Nanomedicine is powerful precisely because nanoparticles do *not* behave like traditional drugs. Nanotoxicology exists because power without understanding becomes a risk.

The Fundamental Shift: From Molecules to Materials

Traditional pharmacology focuses on molecules. Nanomedicine focuses on materials. This is a critical conceptual shift for nonscientific readers. A molecule is like a letter–it carries information. A nanoparticle is like a package–it has shape, size, surface, and structure. Two packages can carry the same letter but arrive very differently depending on how they are wrapped.

Nanotoxicology studies the wrapper, not just the message. This includes how the material interacts with blood proteins; whether immune cells recognize it as friend or threat; how organs such as the liver and spleen process it; and how repeated exposure changes the body's response. In other words, nanotoxicology studies the *relationship* between material and biology over time.

Liposomes Revisited: Familiar Materials, Unfamiliar Behaviors

Liposomes are often introduced as "safe" because they are made from fats similar to those found in our own cells. That is mostly true–but nanotoxicology teaches us that similarity is not identity.

Why Size Still Matters

A fat molecule floating freely behaves differently from a fat molecule arranged into a nanoscale sphere. At very small sizes, liposomes can cross biological barriers, enter immune cells, and circulate longer than free drugs. Nanotoxicology studies whether this persistence is helpful or harmful, depending on context. For example:

- In tuberculosis, persistence inside immune cells is beneficial.
- In vaccines, persistence can amplify immune memory.
- In cancer, persistence may increase efficacy–but also risk.

The Immune System Is Not Passive

One of the most misunderstood aspects of nanomedicine is the immune system's role. The immune system does not simply "allow" materials to pass. It interprets them.

Nanotoxicology examines how liposomes:

- Activate or suppress immune pathways
- Alter cytokine signaling (chemical messengers of immunity)
- Influence inflammation

A liposome that stimulates immunity too strongly may cause fever or inflammation. One that suppresses it may reduce vaccine effectiveness or infection control. Thus, nanotoxicology ensures that liposomes whisper to the immune system—rather than shout.

Iron Oxide Nanoparticles: When Useful Metals Demand Respect

Iron oxide nanoparticles are powerful because they interact with magnetic fields and energy. This makes them useful for imaging, targeted therapy, and cancer treatment. But iron's chemistry is a double-edged sword.

Why Iron Is Both Essential and Dangerous

Iron is necessary for life; it carries oxygen in our blood. But free iron can catalyze reactions that generate reactive oxygen species–molecules that damage proteins, membranes, and DNA. Thus, nanotoxicology asks: Does the nanoparticle release iron uncontrollably? Does it accumulate in organs responsible for iron storage? Does it disrupt normal iron metabolism?

These are not abstract concerns. They are questions of long-term safety, especially for patients receiving repeated treatments.

Coatings Are Not Decoration–They Are Control Systems

Iron oxide nanoparticles are rarely used "naked." They are coated with polymers, sugars, or lipids. Nanotoxicology studies whether these coatings remain intact in the body, whether they degrade safely or they expose the iron core prematurely. This is particularly important in cancer therapy, where repeated dosing may occur.

Polymeric Nanoparticles: When Degradation Becomes Destiny

Polymeric nanoparticles are designed to disappear–slowly. They are meant to release their drug, degrade into harmless byproducts, and exit the body quietly. Nanotoxicology ensures that "quiet" remains quiet.

Why Breakdown Products Matter

Imagine a biodegradable spoon. If it dissolves into sugar, that's fine. If it dissolves into acid, that's a problem. Polymeric nanoparticles degrade into smaller chemical fragments. Nanotoxicology studies whether these fragments alter tissue pH, whether they irritate cells, and whether they accumulate after multiple doses.

This is especially important in vaccines and chronic antimicrobial applications, where exposure may be repeated over time.

Time: The Most Overlooked Variable in Toxicity

One of the most important lessons of nanotoxicology is that time is an active ingredient. A nanoparticle may be harmless today–and problematic after months. Nanotoxicology, therefore, studies:

- Acute effects (immediate)
- Sub-chronic effects (weeks)
- Chronic effects (months to years)

This temporal lens is essential for cancer survivors, patients with chronic infections, and populations receiving repeated vaccination.

Safety is not a snapshot. It is a story.

The Unifying Insight: Biology Does Not Care about Our Intentions

Across AMR, vaccines, and cancer, one truth emerges consistently: Biology responds to interaction, not intention. Nanotoxicology teaches us that healing at the nanoscale requires restraint as much as ingenuity. It reminds us that the same properties that make nanomaterials powerful–size, surface, persistence–also demand accountability.

This does not weaken the case for nanotechnology. It strengthens it. Because technologies that respect biology endure longer, scale wider, and earn trust more easily.

Nanotoxicology and AMR: Avoiding New Problems While Solving Old Ones

As we design nanomaterials to fight resistant microbes, nanotoxicology asks a crucial ethical question: Are we solving resistance–or creating new biological stress? Materials that kill bacteria aggressively may also disrupt beneficial microbiota, stress host tissues, or accumulate in the environment. Nanotoxicology ensures antimicrobial innovation does not repeat the mistakes of indiscriminate antibiotic use.

AMR: When Antimicrobial Innovation Must Not Create New Biological Debt

In the AMR chapters of this book, green-synthesized silver nanoparticles are presented as adaptive, multi-mechanistic tools capable of overwhelming bacterial defenses. Their strength lies in complexity: physical disruption, chemical stress, and biologically programmed surfaces acting together. Nanotoxicology asks a necessary follow-up question:

Can these materials fight microbes without destabilizing host biology or ecosystems?

Unlike classical antibiotics, metallic nanoparticles do not disappear after metabolism. They persist, interact, and transform. In the context of AMR, this persistence can be an advantage–sustained antimicrobial action–but also a risk if accumulation triggers oxidative stress, disrupts microbiota, or contaminates environmental reservoirs.

Crucially, nanotoxicology reframes responsible antimicrobial innovation not as minimizing toxicity alone, but as avoiding the creation of a new resistance-like burden–this time rooted in material exposure rather than microbial genetics. Green synthesis matters here because biologically capped nanoparticles modulate silver ion release, reduce nonspecific cytotoxicity, and align antimicrobial potency with host tolerance. Nanotoxicology thus becomes the *scientific justification* for green nanotechnology–not merely its ethical appeal.

Nanotoxicology and Vaccines: Sustained Immunity Without Sustained Inflammation

Vaccines increasingly rely on nanomaterials to enhance immune responses. Nanotoxicology studies whether immune activation resolves appropriately, whether immune memory forms without exhaustion, and whether inflammatory signals shut down on time. A vaccine that overstimulates immunity may cause harm. One that understimulates may fail. Nanotoxicology helps balance this equation.

In vaccine sections, hydrogels and viral nanoparticles are presented as architectures of sustained immunity–systems that deliver antigens slowly, repeatedly, and in harmony with immune learning. Nanotoxicology provides the framework that makes this strategy credible. Vaccines do not fail because they are insufficiently potent; they fail when immune activation becomes dysregulated. Excess inflammation, unintended biodistribution, or chronic immune stimulation can undermine trust and safety–even when efficacy is high. Here, nanotoxicology focuses less on *what* the material is made of and more on *how long* it stays, where it accumulates, and *which immune pathways* it nudges rather than overwhelms.

Hydrogels, for example, are not inherently safer than soluble injections. Their safety emerges from controlled degradation, local retention, and predictable immune exposure. Viral nanoparticles, likewise, are not dangerous because they are viruses–but because immune systems interpret nanoscale patterns deeply and persistently. Nanotoxicology, in this context, becomes the science of temporal safety–ensuring that vaccines educate immunity without exhausting it.

Nanotoxicology and Cancer: Healing Without Leaving a Scar

Cancer treatment success is no longer measured only by tumor disappearance. It is measured by quality of life after survival. Nanotoxicology ensures:

- Nanoparticles do not accumulate in healthy organs
- Off-target toxicity is minimized
- Long-term complications are avoided

A cure that creates new disease is not progress.

Cancer nanomedicine thrives on selectivity. Targeted delivery, localized heating, immune activation–each strategy depends on nanoparticles behaving exactly as intended. But cancer patients are uniquely vulnerable to unintended effects. Nanotoxicology reveals that materials designed to evade immune clearance, circulate longer, or accumulate in tumors may also persist in the liver, spleen, or bone marrow. Iron oxide nanoparticles, polymeric carriers, and viral platforms all illustrate the same principle: Therapeutic intent does not erase biological consequences.

In cancer therapy, this matters because toxicity is rarely binary. It is cumulative. A nanoparticle that causes no acute harm may still alter immune balance, redox homeostasis, or organ function over repeated cycles. Nanotoxicology thus becomes the discipline that protects the *long arc of survivorship*. It ensures that life-extending therapies do not exchange immediate benefit for delayed harm.

Nanotoxicology as a Translation Filter, Not a Regulatory Obstacle

One of the most damaging misconceptions is that nanotoxicology slows innovation. In reality, it accelerates translation by preventing late-stage failure. Many nanotechnologies stall not because they are ineffective, but because safety questions arise too late, under regulatory scrutiny or public concern. By integrating nanotoxicological thinking early–during material selection, surface engineering, and dosing strategy–developers reduce uncertainty downstream.

This is especially important for low- and middle-income settings, where the cost of failure is not just financial, but societal. Technologies intended to improve global equity must be robust enough to withstand repeated use, variable environments, and long-term exposure.

Nanotoxicology therefore aligns directly with translational justice: it protects communities from becoming testing grounds for poorly understood materials.

Why Nanotoxicology Is a Matter of Global Justice

Nanotoxicology knowledge must circulate as freely as innovation. Without shared safety understanding, low-resource settings may become testing grounds, trust in science may erode, and lifesaving technologies may fail adoption. Equitable science demands equitable safety.

Closing Reflection: Responsibility at the Smallest Scale

Nanotechnology gives us unprecedented power to intervene in biology, to act at the scale of life itself. Nanotoxicology reminds us that power must be accompanied by humility.

Invisible fixes must remain invisible in their harm. Only then can they truly heal the world.

From Precision to Planet: Where Nanotechnology Meets Justice

With green nanotechnology, this book reaches a turning point. Until now, we have focused on engineered precision–how nanoparticles can be designed to deliver drugs, reshape clinical outcomes, and solve problems inside the body. But here the question expands. Nanotechnology becomes not only a matter of performance, but of origin, accessibility, and responsibility. It is no longer only about what nanoparticles can do, but where they come from, what resources and ecosystems they depend on, who can realistically produce them, and whether the benefits of innovation can be distributed beyond a few laboratories, countries, and health systems.

Part IV steps beyond materials and mechanisms to examine the systems that determine impact: scientific leadership, equity in innovation, global collaboration, and the responsibilities that accompany technological power. If the earlier chapters revealed what nanotechnology makes possible, the final section asks how those possibilities can be translated into lasting, just, and global change.

This is where invisible fixes meet visible responsibility.

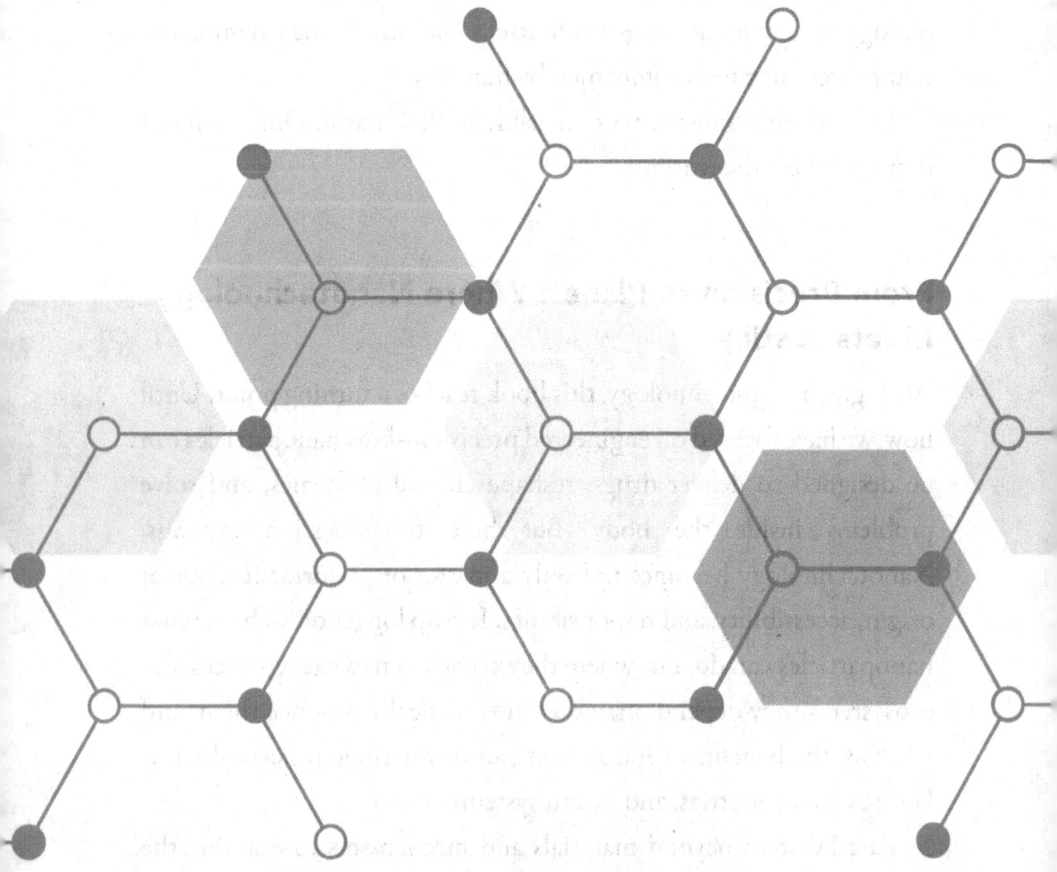

PART IV

THE VISION:

Brain Circulation and Global Health

CHAPTER 14

THE BRAIN DRAIN DILEMMA

Introduction

Brain drain is often measured in numbers–how many professionals leave, how many degrees depart, how much economic value is lost. But its most profound impact is rarely captured in data. It is felt instead in empty laboratories, interrupted mentorships, and generations of students who never meet the teachers they needed.

When a scientist leaves and does not return, the loss is not confined to a single career. It ripples outward–through institutions, through training pipelines, through fragile systems already stretched thin. Knowledge disappears quietly. Experience vanishes without ceremony. And the absence is normalized, even as its consequences accumulate.

Yet mobility itself is not the enemy. Movement, exchange, and exposure are essential to scientific growth. The real failure lies in a global system that too often treats departure as an endpoint rather than a phase–one that extracts talent without building pathways for return, connection, or continuity.

This chapter does not ask whether people should leave or stay. It asks a more urgent question: What kind of science allows knowledge to circulate without hollowing out the places that need it most?

There are moments in life when absence speaks louder than presence. For me, one of those moments arrived quietly, in a university laboratory in Kinshasa.

During my third year of pharmaceutical sciences at the University of Kinshasa, I had a lecturer who shaped my intellectual trajectory in ways I only fully understood years later. He was not a full professor, not a celebrated academic figure whose name appeared on publications or conference programs. He was a senior teaching assistant, responsible for analytical chemistry practical sessions–those hours in the laboratory where theory becomes tactile, where students learn not just equations but judgment, rigor, and discipline.

What distinguished him was not only his mastery of analytical chemistry, but his generosity as a teacher. He explained concepts patiently, connected classroom knowledge to real pharmaceutical quality-control challenges, and spoke with the confidence of someone who practiced science beyond textbooks. At the time, he was also working in quality control for a local pharmaceutical manufacturer, embodying a rare bridge between academia and industry. Watching him teach made analytical chemistry feel alive. It was no longer an abstract subject; it was a tool that protected patients, ensured drug quality, and upheld public trust. Then, without ceremony, he disappeared.

At the start of the following academic year, his voice was gone from the laboratory. His demonstrations were replaced. The rhythm of practical sessions changed. I asked around and eventually learned, through a friend who lived in his neighborhood, that he had left for the United States–and had decided not to return. That was the end of his presence in our faculty.

For the remaining years of my training, I never saw him again. He never returned to teach. He never supervised students. He never passed on what he knew to the next cohort. His departure was not announced as a tragedy. There was no official acknowledgment of loss. But some lab sessions felt incomplete, some students struggled with concepts he would have clarified effortlessly. His absence was felt.

That experience taught me something no statistic ever could: Brain drain is not an abstract migration pattern. It is the quiet erosion of mentorship, continuity, and institutional memory. It is the silence left behind when a teacher does not return.

Brain Drain Beyond Numbers

Brain drain is often described in numbers: how many doctors leave, how many engineers migrate, and how many scientists settle abroad. Those figures matter, but they conceal a deeper truth. What is lost is not merely labor; it is *capacity*. It is the ability of institutions to train the next generation, to sustain standards, to innovate locally, and to respond to crises with homegrown solutions.

As of the mid-2020s, Africa continues to face a widening skills gap. Tens of thousands of highly trained professionals leave the continent every year, drawn by better infrastructure, research funding, salaries, and professional stability elsewhere. Health systems feel this loss acutely. Universities struggle to retain experienced faculty. Research ecosystems remain fragile, dependent on external leadership and short-term projects.

In healthcare alone, the consequences are staggering. Sub-Saharan Africa bears a disproportionate share of the global disease burden while facing a projected shortage of millions of healthcare workers in the coming decade. This mismatch is not simply a function of population growth; it reflects decades of cumulative losses in training capacity, mentorship, and institutional reinforcement.

The economic cost is equally severe. When a country invests in educating a pharmacist, physician, or scientist who later leaves permanently, that investment benefits another system. Over time, this dynamic contributes to structural dependency, where solutions must be imported rather than developed locally.

Yet, despite its devastating impact, brain drain is rarely discussed with nuance. It is often framed as a moral failure–of individuals who "abandon" their countries–or as an inevitable outcome of globalization. Both framings are incomplete.

The Temptation to Leave

I intimately understand the pull of departure.

In 2017, while I was a doctoral student in South Africa, I traveled to Europe for the first time. The contrast was overwhelming. Well-equipped laboratories. Predictable funding mechanisms. Seamless access to reagents. Institutions designed to support research rather than merely tolerate it. Each visit–to the United Kingdom, Germany, Switzerland–came with conversations that followed a familiar script.

"Why go back?"
"You could stay here."
"You would have a better life."

These suggestions were not malicious. They were offered as advice, sometimes as concern, sometimes as admiration. And they were repeated at every international meeting I attended: antimicrobial resistance workshops, nanotechnology conferences, industry-academic programs.

I resisted–not because the temptation was absent, but because at that stage of my life, returning was not optional. I still had commitments. I still had unfinished training. But I often wondered: *What if I had been at a different point? What if I had no clear path home?* Many do not.

For countless scientists, clinicians, and engineers from low- and middle-income countries, staying abroad is not a betrayal–it is a survival strategy. Stable salaries, access to healthcare, educational opportunities for children, and professional recognition are powerful incentives. Blaming individuals for responding rationally to structural inequities misses the point entirely.

Rethinking the Narrative: From Drain to Circulation

Over the past two decades, scholars have increasingly challenged the simplistic notion of brain drain as an irreversible loss. Alternative frameworks have emerged: brain gain, brain banks, brain bridges, and most compellingly, brain circulation.

These perspectives recognize that mobility does not automatically sever ties. Scientists living abroad often remain connected to their countries of origin through collaborations, mentorship, co-authorship, funding facilitation, and technology transfer. In some cases, their presence abroad amplifies their ability to support colleagues back home.

Empirical studies have demonstrated that researchers who collaborate with diaspora scientists publish more internationally, access broader networks, and increase their scientific visibility. Diaspora scientists act as connectors–translating standards, expectations, and opportunities across borders.

This reframing does not deny loss. Instead, it asks a more productive question: Under what conditions does mobility become an asset rather than a deficit?

The answer lies not in forcing people to return permanently, but in creating systems that value circulation–movement with continuity, exchange without erasure, return without any challenges.

When Return Is Possible–and When It Is Not

Return is often romanticized. But returning home after years abroad is not a simple reversal. Scientists who attempt it frequently face underfunded institutions, limited infrastructure, bureaucratic barriers, and professional isolation. Without deliberate reintegration mechanisms, return can become professionally damaging rather than empowering.

Short-term visits, joint appointments, visiting professorships, and co-supervision models offer more realistic pathways. Virtual collaboration platforms, shared grant mechanisms, and regional centers of excellence can further reduce the burden of physical relocation while preserving impact.

Some governmental and international initiatives have recognized this reality, engaging diaspora professionals as partners rather than lost assets. These efforts, while commendable, often remain modest relative to the scale of the challenge.

What is still missing, particularly in the health sciences, is a deliberate alignment between local needs and global expertise–not merely for publications, but for capacity building, product development, and system resilience.

My Own Trajectory as a Case Study

Looking back over the last two decades, my journey has taken me from Mbandaka to Kinshasa, from South Africa to Europe, from academic laboratories to industry-aligned research in the United States–and back again.

Each stage expanded my scientific vocabulary. Each environment sharpened different skills. But none replaced the responsibility I felt toward the systems that shaped me.

Returning to Kinshasa as an active researcher was not an act of nostalgia. It was a strategic choice–one rooted in the conviction that global science only fulfills its promise when it strengthens, rather than bypasses, the places most affected by disease, scarcity, and environmental stress.

As presented in the next chapter, I thrived to establish research programs, mentor students, build laboratories, and link local challenges to global technologies. These efforts are neither heroic nor sufficient on their own. But they illustrate a principle: Circulation is intentional. It does not happen by accident. It must be cultivated, supported, and protected.

Brain Drain as a Health Issue

In the context of this book, brain drain is not merely a development issue; it is a health issue.

Pandemics, antimicrobial resistance, climate-driven disease patterns, and emerging infections do not respect borders. A shortage of trained professionals in one region increases vulnerability everywhere. Weak surveillance systems, under-resourced laboratories, and limited pharmaceutical capacity create blind spots in global preparedness.

If nanotechnology, biotechnology, and pharmaceutical innovation are to serve humanity equitably, the knowledge required to deploy them must circulate as well. Invisible fixes fail when the people needed to implement them are absent.

This chapter does not argue that everyone must return home. It argues something more demanding: that global science must be designed to allow meaningful participation from everywhere, not just from where resources are concentrated.

Toward the Next Chapter

Brain drain, then, is not the end of the story. It is the tension that forces a deeper reimagining of how science moves, who it serves, and where it takes root.

In the next chapter, I shift from diagnosis to possibility. From loss to motion. From absence to connection.

Brain circulation–when done deliberately, ethically, and inclusively–offers a path forward. A model of global science that does not extract talent but multiplies it. That does not demand permanent return but enables lasting impact.

That is where the vision begins.

CHAPTER 15

BRAIN CIRCULATION AND GLOBAL SCIENCE AS AN INVISIBLE FIX

Introduction

Science moves with people. When people leave, something leaves with them–not just skills or diplomas, but mentorship, memory, and momentum. Yet when movement is guided by intention rather than escape, mobility can become a source of renewal rather than loss. This chapter explores how scientific migration, long framed as a dilemma, can instead become an invisible fix–one that transforms distance into connection, absence into circulation, and global science into a shared responsibility.

"I would like to travel and do science abroad, because they have all the infrastructure needed for research."

That sentence, spoken with honesty and youthful urgency, followed me longer than I anticipated. I gave it as my answer to the selection panel of the BEBUC Excellence Scholarship at the University of Kinshasa in 2010, when I first applied as an undergraduate student.

I believed it fully. At that moment in my life, it felt self-evident: Meaningful science could only be done elsewhere, in places where equipment was functional, reagents available, and laboratories alive. Kinshasa, to my eyes then, was a place of intellectual potential constrained by structural absence.

When my application was unsuccessful, the disappointment felt terminal. I had interpreted the opportunity as providential–almost divinely ordained. I was a strong student, disciplined, curious, and committed. I searched for explanations and found none that satisfied me. It was my late brother, Corentin Nkanga, who named the uncomfortable truth with clarity and kindness: "They probably felt that you wanted to go abroad and never return."

At the time, he was right. I could not honestly imagine a future in which I could thrive scientifically in Kinshasa without external support. I did not yet see returning home as a responsibility; I saw it as a limitation. What changed later was not merely my language, but my understanding of what science could mean–and where it could live.

As I explained in chapter 2, when I reapplied five years later, I was awarded the BEBUC Master's Scholarship, I spoke differently. I spoke of learning abroad and returning home to teach. I signed the terms committing myself to return. Yet I must be honest: That conviction was still fragile. It was more contractual than philosophical. What truly reshaped my thinking came later, through experience, frustration, loss, and ultimately, responsibility.

From Brain Drain to Brain Bridge

During my doctoral training at Rhodes University in South Africa, I began–almost unintentionally–to practice what would later become central to my scientific identity: acting as a *brain bridge*. Even as a student, I refused to sever my intellectual ties with the University of Kinshasa. Collaborations emerged organically, particularly with Professors Patrick B. Memvanga and Kalulu M. Taba.

Through these collaborations, I supported research projects for several students focusing on self-nanoemulsifying drug-delivery systems, silica nanoparticles, green synthesis of metallic nanoparticles from Congolese plants, etc. Plant materials were sent to me; experiments were conducted in South Africa with permission from my late doctoral supervisor, Professor Rui W. M. Krause; data and interpretations flowed back to Kinshasa. This arrangement produced several dissertations. It produced publications. But it also produced discomfort.

Students received results without touch. Knowledge moved, but hands did not. Training occurred at a distance, filtered through explanations rather than experience. Even as a bridge, I could see the limitations of a one-directional flow. Brain bridges without local infrastructure risk becoming symbolic rather than transformative.

Returning Home–and Facing the Gap

When I returned to the University of Kinshasa in 2019 as an associate professor, teaching pharmaceutical chemistry, nanomaterials, ethnopharmaceutics, artificial intelligence, and scientific English, the joy of teaching was real–but so was the frustration. I was mentoring my first nanotechnology student, full of enthusiasm, yet we had no particle sizer, no zeta potential analyzer, no spectroscopy tools.

Students mixed emulsions without knowing whether nanoparticles had formed. Samples had to be sent abroad. Again.

At that point, the tension became unbearable. My curiosity was not exhausted. My sense of responsibility was growing, but the tools were absent. And then, in a moment that captures the unpredictability of modern science, opportunity arrived through a social media post.

Mobility with Intention

In September 2019, Professor Karine Ndjoko tagged me in a post on X announcing a postdoctoral opening at the University of California San Diego. I applied. Weeks later, I was selected. Professor Ndjoko and I happened to have a conference at the University Kongo, and during our road trip, I informed her of the outcome. She smiled–and then told another colleague travelling with us, half-jokingly, half-fearfully, "The United States will take him away from us." I smiled too. But I already knew something she did not yet know: I was not leaving to disappear.

In January 2020, as I prepared to travel to San Diego, Corentin died from untreated liver disease in Kinshasa. The timing was cruel. The grief was disorienting. But the decision remained unchanged. I left not to escape, but to learn faster, deeper, and more deliberately–because the absence of therapeutic answers had become personal.

Circulation Can Be Made Tangible

Two years later, after intense postdoctoral training, I returned to Kinshasa with the ARISE Fellowship. This time, I returned with more than ideas. I returned with resources, partnerships, and a mandate.

The nanotechnology laboratory I established under ARISE

transformed everything. Students no longer watched science from afar; they performed it. They synthesized nanoparticles, measured sizes, analyzed surfaces, interpreted spectra. At one point, one of my students visiting Rhodes University ran out of particle analyzer consumables and requested from Kinshasa. We shipped it to them. That moment mattered more than any publication; the flow had reversed.

Several graduate students from my research programs presented their work internationally. They published. They competed. They belonged.

What followed was not accidental. International collaborations multiplied–not because of charity, but because quality attracts attention. Grants followed. Partnerships deepened. From Tanzania to Canada to the United Kingdom, networks expanded. In December 2024, the *Centre de Recherche en Nanotechnologies Appliquées aux Produits Naturels* (CReNAPN) was officially launched at the University of Kinshasa.

This was no longer brain circulation in theory. It was circulation embodied—in equipment, in students, in institutions.

A Call Beyond One Story

My journey does not solve the global problem of inequity in science. It does not bring back my brother. It does not erase structural injustice. But it demonstrates something essential: *Mobility does not have to mean loss*. Knowledge can move without hollowing out its origins. Excellence can be global without being extractive.

Brain circulation is not a slogan. It is a practice–one that requires intention, infrastructure, trust, and accountability. Scientists from low-income settings do not need to choose between staying invisible or leaving permanently. A third path exists: travel to learn, return to build, and circulate continuously.

This chapter is written as an invitation. Not to copy a trajectory, but to imagine responsibility differently. The invisible fix to global scientific inequality is not immobility. It is purposeful circulation.

The final chapter builds on this idea, asking the most difficult question of all: *How do we design systems that make equitable science the norm rather than the exception?*

CHAPTER 16
BUILDING EQUITABLE SCIENCE

Introduction

Scientific excellence without equity is not innovation–it is imbalance. When research capacity, funding, infrastructure, and voice are concentrated in a few regions of the world, global science becomes structurally fragile, incapable of responding fairly or effectively to shared threats. Building equitable science therefore demands more than individual brilliance or isolated programs; it requires intentional policy choices, sustained investment in local capacity, and governance frameworks that recognize knowledge as a global public good.

This chapter argues that equity in science must be treated as infrastructure: designed, financed, maintained, and evaluated with the same seriousness as laboratories, hospitals, or power grids. Without policies that enable access, reward mentorship, support mobility with return, and institutionalize inclusion, even the most promising scientific breakthroughs risk reinforcing the very inequalities they aim to solve.

Equitable science is not charity, and it is not symbolism. It is the deliberate construction of systems that ensure fairness, justice, and genuine opportunity in the production of knowledge, the training of scientists, and the distribution of scientific benefits. At its core, equitable science recognizes that talent is universal, but opportunity is not–and that this imbalance is neither natural nor inevitable. It is the result of historical structures, policy choices, and institutional priorities. Correcting it requires intention, persistence, and courage.

Equitable science applies to everyone, irrespective of race, gender, geography, income, or background. It is not a concession to the disadvantaged; it is a prerequisite for global progress. In a world facing pandemics, climate instability, antimicrobial resistance, food insecurity, and environmental degradation, science that excludes large portions of humanity is not only unjust–it is ineffective.

My own commitment to equitable science did not emerge from theory. It was forged through experience–through moments of inspiration, exclusion, mentorship, and service. During my postgraduate years at Rhodes University in South Africa, I became deeply engaged in community outreach and science engagement, an ethos I inherited directly from my late supervisor, Professor Rui W. M. Krause. Professor Krause believed that science did not belong exclusively in laboratories, journals, or conferences; it belonged in classrooms, town halls, and communities.

Through his research group, postgraduate students were encouraged–and expected–to engage with society. We spent countless hours working with high school learners and teachers through the Khanya Maths and Science Club, helping students navigate homework, experiments, and confidence. We participated actively in the Africa Science Festival (SciFest), where science ceased to be abstract and became tangible, playful, and human.

I vividly remember standing beside Professor Krause during chemistry demonstrations, guiding rotating groups of learners through experiments. One demonstration in particular–the rapid formation of colorful gels–became my favorite. It was not because of the chemistry alone, but because of the learners' reactions: wide eyes, laughter, questions, and that unmistakable moment when curiosity ignites. I was often so busy preparing, cleaning, resetting, and engaging that Professor Krause once joked, "Christian, you would make a great restaurant man," a comment that still makes me smile. What he truly meant was that service, attentiveness, and joy matter in science.

That period planted a lasting conviction in me: Science that does not engage society cannot claim to serve it.

Christian I. Nkanga (far left) with High School learners during the Khanya Maths and Science Club meeting at the Albany Museum of Rhodes University, Grahamstown, South Africa. 2018. Photo credit: Alain M. Bapolisi.

From Community Engagement to Scientific Citizenship

My involvement in outreach gradually expanded my interest beyond laboratories into science academies and scientific citizenship. When one of my BEBUC mentors, Professor Karine Ndjoko, proposed the idea of creating a national young academy of sciences in the Democratic Republic of the Congo, I immediately recognized its significance. This was not merely about prestige or representation; it was about voice.

Together with colleagues Tania T. Bishola, Freddy F. Bangelesa, Hervé M. Lekuya, Alain M. Bapolisi, George M. Kasay, Florent Zeng, Salomé Kahindo, Blaise Pascal Zirim, and Félicien Mushamuka, we embarked on a demanding two-year journey to establish what would become the Académie des Sciences pour les Jeunes en République Démocratique du Congo (ASJ-RDC), officially launched in November 2020.

All ten members of the founding committee were BEBUC scholars, bound not only by shared academic experience but by a collective sense of responsibility. I was honored to serve as the committee's president, with Tania as vice president, but leadership in this endeavor was deeply collective. Every member contributed tirelessly—drafting statutes, defining selection criteria, consulting stakeholders, and building legitimacy from the ground up.

Our vision was clear: ASJ-RDC would be the voice of young Congolese scientists, both within the country and across the diaspora. Guided by the motto "*Science au service de la nation*" –science in service of society–we sought to identify and unite Congolese PhD holders worldwide through a competitive, merit-based process. But recognition was only the beginning.

Equitable science, by definition, requires action. ASJ-RDC rapidly developed initiatives to support emerging scholars, including:

- Providing free editorial and manuscript-review support to improve publication quality
- Creating platforms for outstanding young researchers to present their work through monthly webinars of the academy
- Offering mentorship and visibility to scholars who otherwise remained unseen

Crucially, we also committed to open science outreach. Monthly public webinars addressed pressing issues such as mpox, flooding, climate change, and vaccines–topics of immediate relevance to Congolese society. Science was no longer confined to elite spaces; it was translated, contextualized, and shared.

Reaching the Next Generation

Equitable science must be intergenerational. Drawing directly from my Rhodes University experience, I began engaging with high school learners in the DRC. In November 2023, I visited Mont Amba School in Lemba Township, speaking to more than three hundred students about science, curiosity, and possibility. That visit marked the beginning of a broader movement. Other ASJ-RDC members soon followed, visiting schools across the country.

At the university level, Professor Paulin Mutwale introduced the Nanotechnology Day concept, creating an annual platform to demystify nanoscience and nanotechnology for students, academics, policymakers, and the media. Over the years, these conferences brought together national and international speakers–including colleagues from South Africa–and helped embed nanotechnology into public discourse in the DRC.

Science communication extended beyond academic halls. Over time, I was invited by journalist Hervé Pedro to discuss nanotechnology on *Ecosystemic*, a television program on EDUCTV. Speaking to a national audience reinforced an essential truth: Visibility matters. A science that is not seen cannot inspire.

From National to Global Advocacy

Equitable science does not stop at borders. In August 2023, my colleague and ASJ-RDC cofounder, Dr. Hervé Lekuya, encouraged me to apply for membership in the Global Young Academy (GYA). I already shared that aspiration, and with the recommendation support of Professor Kalulu M. Taba, I was selected in May 2024 as one of forty-five new members worldwide–the first Congolese scientist to ever join the GYA.

This milestone was deeply symbolic, but more importantly, it was functional. I joined the GYA not for personal recognition, but to amplify advocacy for equitable science on a global platform. I have always believed that exposure to science can awaken dormant potential. As someone who grew up in Mbandaka, in a family where only three out of eleven children attained a bachelor's degree, I often remind young people that a disadvantaged background does not nullify dreams.

Within my first year at the GYA, I initiated several advocacy initiatives, including a webinar series demystifying the GYA membership application process–an effort aimed at widening access and countering informational inequity. The initiative was later institutionalized by the academy.

In my second year, I was elected to the GYA Executive Committee, one of only eleven members globally–an unexpected but powerful affirmation of trust. This election, which occurred in my absence from the annual meeting, reflected the collective recognition of sustained engagement rather than visibility alone.

Institutional Change and Capacity-Building

My advocacy work opened doors beyond membership academies. In July 2024, I received an invitation from the UK Academy of Medical Sciences (AMS) to join the steering committee of an African Research Capacity Strengthening Workshop, cohosted with the Academy of Sciences of South Africa and the Network of African Science Academies (NASAC).

Working alongside distinguished scholars, including Professor Assan Jaye, we designed a program for sixty outstanding early-career African researchers from twenty-eight African countries. The workshop, held in Johannesburg in November 2024, focused on grant writing, leadership, collaboration, and scientific communication.

Balancing this role with my responsibilities at the University of Kinshasa, ASJ-RDC, GYA, and my research programs was demanding. Yet witnessing the transformation–confidence gained, collaborations formed, ambition sharpened–made every effort worthwhile. This was equitable science in action; not rhetoric, but capacity built.

Illustrative Prescriptions for Building Equitable Science

From these experiences, several prescriptions can clearly emerge:

1. **Governments must invest in people and infrastructure simultaneously.** Talent without tools leads to frustration; tools without talent lead to stagnation.

2. **Diaspora engagement should be structured, not symbolic.** Brain circulation requires funded pathways for return, collaboration, and co-ownership.

3. **Science academies must be empowered as independent voices,** not ceremonial bodies.

4. **Community engagement must be institutionalized,** not left to individual goodwill.

5. **Global partnerships must be reciprocal,** recognizing expertise in the Global South as contributory, not merely extractive.

6. **Young scientists must be visible early,** because representation fuels aspiration.

Equitable science is built deliberately–through policies, platforms, mentorship, and sustained commitment.

A Personal Closing Reflection

In November 2024, shortly after the AMS workshop in South Africa, I went to visit Rhodes University, following the passing of my graduate supervisor, Professor Rui Krause. With the help of my students still based there, I spent time with his wife and children, remembering a man whose legacy continues to shape my values. That moment reminded me that science is ultimately human–carried through relationships, memory, and responsibility.

This chapter, and this book, are therefore not conclusions. They are invitations. A call to scientists to circulate rather than disappear. To institutions to invest rather than exclude. And to societies to recognize that invisible fixes–when nurtured–can reshape the visible world.

EPILOGUE – MAKING THE INVISIBLE INEVITABLE

This book began with absence–with a question born from loss, pain, and silence. It ends with responsibility.

Across these pages, I have explored technologies too small to see, systems too fragile to ignore, and choices too consequential to postpone. In Part III, I translated some of my own published research into narrative–not to simplify science, but to make it legible, human, and ethically urgent. Nanotechnology emerged not as a miracle cure, but as a lens–one that reveals how matter, medicine, and society interact at their most fundamental levels. From liposomes engineered to outsmart tuberculosis, to plant viruses redesigned as cancer immunotherapies, to hydrogels that turn time into a vaccine ingredient, to green silver nanoparticles confronting antimicrobial resistance, the message is consistent: The nanoscale is not a gimmick–it is a design space. Brain circulation appeared not as an abstract theory, but as a lived strategy for restoring balance between places of abundance and places of need. Equitable science revealed itself not as charity, but as design. And nanotoxicology–often ignored in popular narratives–reminded us that every new platform carries trade-offs. Progress is not only invention; it is also restraint, testing, transparency, and long-term responsibility.

The central message is simple yet demanding: Solutions already exist–but they remain invisible not because they are undeveloped, but because translation systems have failed to carry them where they are needed most. We live in a world where scientific capacity is unprecedented, yet access remains uneven. Where cures are engineered faster than they are distributed. Where talent flows freely across borders, but opportunity does not. These are not failures of science. They are failures of structure, policy, and collective will. They are also failures of translation: the long chain of documentation, manufacturing, regulation, clinical adoption, and financing that turns a promising experiment into a medicine someone can actually receive.

The invisible fixes described in this book–nanomedicines, liposomes, virus-based nanomedicines, hydrogels for sustained immunity, green nanoengineering against AMR, long-acting therapies, nanotoxicology as a safety compass, global scientific mobility, diaspora bridges, young academies, and community-rooted research–share a defining trait: They work best when science is treated as a shared responsibility–not a gated advantage. They demand ecosystems where knowledge circulates, where infrastructure follows talent home, where innovation is measured not only by citation counts or market size, but by lives reached and systems strengthened. This is not a call for idealism detached from reality. It is a call for pragmatic courage.

For policymakers, it means embedding equity into research funding models, regulatory frameworks, and innovation pipelines–so that translation does not stop at borders. This includes regional regulatory strengthening, harmonized pathways for complex nanomedicines, and incentives that reward translation for neglected diseases–not only blockbuster markets.

For funders, it means shifting from extractive excellence to sustained partnership, investing in people and platforms, not just projects. It also means funding the unglamorous middle: quality systems, GMP readiness, biostatistics, clinical partnerships, and local manufacturing feasibility–so that discoveries do not die in pilot purgatory. For institutions, it means valuing return, mentorship, and capacity-building as much as mobility. Promotion and recognition systems must reward translation and capacity-building–not only papers, impact factors, and overseas affiliations. For scientists, especially those who move between worlds, it means embracing the responsibility of connection–serving as bridges, not endpoints. It also means learning the language of translation: documentation, reproducibility, quality, regulation, and the realities of the clinic.

And for young minds reading these pages–those who doubt whether their background limits their future–this book stands as evidence that origin does not define destiny, but commitment defines impact.

Invisible fixes become inevitable only when systems are built to sustain them. When science is allowed to flow back, not just outward. When innovation is designed with inclusion, not retrofitted after exclusion. When resilience is recognized as a form of intelligence.

The work does not end here. This book is not a conclusion–it is a handoff.

To the reader: Whether you are a researcher, clinician, policymaker, student, or advocate, you are already part of the solution. The question is not whether the tools exist. The question is whether we will choose to align them–with fairness, with foresight, and with urgency.

Because in a world bound by shared vulnerability, equitable science is no longer optional.

It is the only future that does not collapse. Cancer, pandemics, and antimicrobial resistance are different crises, but they share one truth: biology travels faster than equity.

ACKNOWLEDGEMENTS

There are so many people to whom I am deeply grateful–people without whom this book would not have been possible.

First, my heartfelt thanks to my wife, Suzanne M. Bigohe. During the writing of this book, my computer often became a rival–"another wife," as she would jokingly say. Your patience, understanding, and quiet strength carried me through the long hours and many absences.

To my daughters, Yemi and Zasti: you gave this book its most human dimension. There were many evenings when you came to my office, hoping to pull me away from my screen and into your world. Too often, you fell asleep waiting. Many mornings, you woke up to find me still there, typing, as if the night had never passed. Your love, even in those moments of distance, sustained me more than you know.

I am deeply grateful to the editorial team at Reedsy who helped bring this book to life: Joe Pierson for his thoughtful developmental and copyediting support; Louise Brigenshaw for her creativity and care in design; and Wayt Gibbs for his valuable discussions and some guidance throughout the process.

My sincere thanks to my brother, Junior Bomolo, for helping me revisit and validate the memories of our childhood that shaped so much of this story.

I am especially grateful to Professor Twan Lammers for his careful fact-checking, his thoughtful review of the manuscript, and for generously writing the foreword despite his demanding schedule.

I also extend my appreciation to my colleagues at the Centre de Recherche en Nanotechnologies Appliquées aux Produits Naturels (CReNAPN), as well as to my friends, collaborators, mentors and those who taught/trained me at the Institut Frere Iloo, University of Kinshasa, Rhodes University, and the University of California, San Diego. Each of you has contributed, in different ways, to the journey that made this book possible.

Finally, to all those I may not have named but who have supported, encouraged, or inspired me along the way–please know that I am deeply grateful.

Thank you.

GLOSSARY:

Understanding the Language of Invisible Fixes

A Mini-Dictionary for Invisible Fixes

A

Access (to Innovation / Healthcare)

Access refers to the real, practical ability of individuals or communities to benefit from scientific or medical advances. Access is not determined only by whether a technology exists, but by affordability, infrastructure, supply chains, regulatory approval, trained personnel, and geographic availability. A medicine may be scientifically validated yet remain inaccessible to most of the world.

Accumulation

The gradual build-up of a substance in tissues over time. In nanotoxicology, accumulation matters because some nanoparticles persist longer than conventional drugs, raising concerns about long-term exposure, even at low doses.

Active Pharmaceutical Ingredient (API)

The active pharmaceutical ingredient is the chemical or biological substance in a medicine that produces the intended therapeutic effect. Everything else in a drug formulation exists to support the API–by stabilizing it, helping it dissolve, improving absorption, or controlling how it is released. In nanomedicine, APIs are often encapsulated inside nanoparticles to overcome problems such as poor solubility, rapid degradation, or high toxicity.

Aggregation

Aggregation occurs when nanoparticles clump together instead of remaining evenly dispersed. This can change how they behave in the body, reduce effectiveness, block blood vessels, or increase toxicity. Preventing aggregation is a central challenge in nanoparticle formulation and scale-up.

Analytical Characterization

Analytical characterization is the process of measuring and confirming the physical, chemical, and biological properties of a material. For nanomaterials, this includes size, shape, surface charge, composition, stability, and drug content. Without rigorous characterization, scientific results cannot be reproduced or translated into real products.

Antibiotic Pipeline

The process through which new antibiotics are discovered, tested, approved, and made available for clinical use. A "pipeline crisis" refers to the sharp decline in the number of new antibiotics being developed, largely due to high costs, regulatory hurdles, and limited commercial incentives, leaving society vulnerable to resistant infections.

Antigen

An antigen is any substance–often a protein or a small piece of a virus, bacterium, or tumor cell–that the immune system recognizes as "not me." When your body detects an antigen, it triggers an immune response and begins building defenses (like antibodies and immune memory), so it can respond faster if it encounters the same threat again.

Antimicrobial Resistance (AMR)

Antimicrobial resistance is the ability of microorganisms to survive treatments that previously killed them. AMR threatens modern medicine by making infections harder–or impossible–to treat. Nanotechnology offers alternative antimicrobial mechanisms that do not rely on traditional antibiotics.

B

Bench Science

Bench science refers to laboratory-based experimental research, typically performed at small scale. While essential for discovery, bench science alone does not guarantee real-world impact unless paired with translational development.

Bench-to-Bedside

Bench-to-bedside describes the full journey from laboratory discovery to patient treatment. This pathway includes formulation development, safety testing, regulatory approval, manufacturing, clinical adoption, and monitoring. Many innovations fail between bench and bedside due to missing translational capacity.

Biovailability

Bioavailability is the fraction of a drug that successfully reaches systemic circulation and becomes available to act in the body. Nanoparticles can improve bioavailability by protecting drugs from degradation and improving absorption.

Biocompatibility

A measure of how well a material can exist in the body without causing harmful immune, inflammatory, or toxic responses. In nanomedicine, biocompatibility is not absolute–it depends on dose, route of administration, surface chemistry, and exposure time. A nanoparticle may be biocompatible in blood but irritating in lung tissue.

Biodegradability

The ability of a material to break down into nontoxic components within the body. Biodegradable nanoparticles (such as polymeric carriers) are generally preferred for chronic treatments, as they reduce accumulation risk.

Biodistribution

The pattern by which a substance travels through and accumulates in different organs and tissues after administration. Nanoparticles often distribute differently than conventional drugs because of their size. For example, many nanoparticles naturally accumulate in the liver and spleen, which are part of the body's filtration system.

Biofilm

A structured community of microorganisms embedded in a self-produced protective matrix and attached to surfaces such as wounds, medical devices, or tissues. Biofilms make infections difficult to treat because they block antibiotics, reduce drug penetration, and allow microbes to survive in dormant states.

Biohybrid Nanomaterial

A nanoscale material composed of both biological components (such as plant phytochemicals, proteins, or viral particles) and inorganic elements (such as metals). In green nanotechnology, silver nanoparticles capped with plant molecules are examples of biohybrid systems that combine biological function with physical stability.

Brain Circulation

Brain circulation is the dynamic movement of knowledge, skills, and people across borders, combined with ongoing collaboration and return. Unlike brain drain, it strengthens both host and home countries by enabling shared growth.

Brain Drain

Brain drain refers to the permanent loss of skilled professionals from low-resource settings to high-income regions, often weakening local institutions and capacity.

C

Capacity-Building

Capacity-building involves strengthening human expertise, infrastructure, institutions, and systems to enable sustained scientific and technological progress. It includes training researchers, establishing laboratories, developing regulatory expertise, and building manufacturing and translational skills.

Capping Agents

Molecules that coat nanoparticles and stabilize them. In green nanotechnology, capping agents are often plant-derived compounds that influence toxicity, antimicrobial action, and biological interactions.

Clearance

The process by which a substance is removed from the body. Nanoparticles can be cleared through renal clearance (kidneys, typically for very small particles), hepatobiliary clearance (liver and bile), and immune-mediated clearance (uptake by macrophages). Clearance rate strongly influences both efficacy and safety.

Clinical Translation

Clinical translation is the process of converting a scientific discovery into a therapy that can be tested and used in patients. It requires expertise beyond basic science, including regulatory science, manufacturing, clinical trial design, and quality systems.

Contract Development and Manufacturing Organization (CDMO)

A CDMO is a specialized company that helps develop, scale, and manufacture pharmaceutical products. CDMOs bridge the gap between laboratory research and clinical or commercial production, offering expertise that academic labs and startups often lack.

Controlled Release

Controlled release refers to the engineered, predictable release of a drug over time, rather than immediate delivery. This is achieved using materials such as biodegradable polymers, liposomes, hydrogels, solid lipid nanoparticles, or long-acting injectable depots. Controlled release can occur through material degradation, diffusion, environmental triggers (pH, enzymes), or mechanical erosion. For example, a polymeric nanoparticle may slowly degrade in the body, releasing its drug payload over weeks, reducing dosing frequency and side effects.

D

Data Integrity

Data integrity ensures that research data are accurate, complete, traceable, and reliable. Poor data integrity leads to irreproducible research and failed translation.

Decentralized Manufacturing

A production model in which medical or technological products are manufactured locally or regionally, rather than in a few centralized global facilities. In the context of green nanotechnology, decentralized manufacturing allows communities to produce antimicrobial materials using local plant resources, improving access and resilience.

Development Pipeline

The development pipeline refers to the sequential stages required to turn a discovery into a usable product. For medical innovations, this pipeline includes discovery, formulation, preclinical testing, clinical trials, regulatory approval, manufacturing, and post-market surveillance.

Dose–Response Relationship

The relationship between how much of a material is administered and the biological effect it produces. Nanomaterials can show nonlinear dose–response curves, meaning small dose increases can sometimes produce disproportionate effects.

E

Efflux Pump

A biological mechanism used by bacteria to expel antibiotics from their cells. Efflux pumps reduce drug accumulation inside microbes and are a major contributor to antimicrobial resistance.

Encapsulation

Encapsulation is the process of enclosing a drug inside another material–such as a nanoparticle, liposome, or polymer matrix–to protect it, improve delivery, and control release.

Environmental Nanotoxicology

The study of how nanomaterials affect ecosystems, soil, water, and nonhuman organisms. This is particularly relevant for antimicrobial nanoparticles that may enter the environment after use.

Equitable Science

Equitable science ensures that scientific benefits, opportunities, and resources are distributed fairly across populations and regions, not concentrated only where science is produced.

Exposure Route

The pathway through which a substance enters the body, such as intravenous injection, inhalation, oral ingestion, or skin contact. Nanotoxicity depends strongly on exposure route. For example, nanoparticles safe in blood may be harmful if inhaled.

F

Failure Rate

Failure rate refers to the proportion of scientific projects, startups, or technologies that do not successfully reach real-world application. In health technology, failure rates are high–often from 60 to 80 percent–due to regulatory, clinical, manufacturing, or reimbursement barriers. High failure rates highlight the need for stronger translational capacity.

Formulation Science

Formulation science studies how active ingredients are combined with supporting materials to create stable, effective products. In nanomedicine, formulation determines particle size, stability, release behavior, and safety.

Functional Nanomaterial

A functional nanomaterial is a nanoscale material engineered to perform a specific task, such as targeting cancer cells, releasing drugs, killing bacteria, or sensing radiation. Functionality arises from size, surface chemistry, structure, or responsiveness to environmental conditions.

G

Good Documentation Practices (GDP)

GDP refers to standardized methods for recording experimental procedures, data, and decisions clearly and accurately. Poor documentation is a major cause of irreproducible research and failed translation.

Good Manufacturing Practice (GMP)

GMP is a regulatory framework ensuring that products are consistently produced and controlled according to quality standards. Research-grade materials often behave differently from GMP-grade materials, requiring re-optimization during scale-up.

Green Nanotechnology

Green nanotechnology uses environmentally friendly methods and biological materials–such as plant extracts or waste–to produce nanoparticles. It reduces toxic solvents, lowers costs, and supports sustainability and local production.

Green Synthesis

The production of nanoparticles using biological systems such as plants or microbes, avoiding toxic chemicals. Green synthesis often enhances biocompatibility by adding natural surface coatings.

H

Health Equity

Health equity means ensuring that everyone has a fair opportunity to achieve good health. It requires addressing structural barriers, unequal access, and systemic disadvantages.

Hydrogel

A soft, water-rich polymer network capable of holding large amounts of fluid. Hydrogels can encapsulate drugs, nanoparticles, or viral particles and release them slowly over time. In immunotherapy and vaccination, hydrogels act as "depots" that prolong immune stimulation.

I

Immune Recognition

The process by which the immune system detects and responds to foreign materials. Nanoparticles may be ignored (stealth behavior), activate immunity (useful for vaccines), or trigger inflammation (undesirable toxicity). Designing nanomaterials often involves balancing visibility versus invisibility to immune cells.

Immunogenicity

The ability of a substance to provoke an immune response. In vaccines, immunogenicity is desired. In drug delivery, unintended immunogenicity can cause allergic reactions, inflammation, or rapid clearance. Nanotoxicology evaluates when immune activation is helpful versus harmful.

Immunotherapy

A treatment approach that harnesses the body's immune system to fight disease, particularly cancer. Nanotechnology can enhance immunotherapy by delivering immune-stimulating agents precisely and sustainably.

Implementation Gap

The implementation gap is the disconnect between scientific discovery and real-world use. Many technologies work scientifically but fail to be implemented due to missing infrastructure, policy, or funding.

Iron Oxide Nanoparticles

Magnetic nanoparticles often used for imaging, drug delivery, or hyperthermia. Their nanotoxicology depends on iron release, oxidative stress, and immune interactions. Proper coating and dosing are critical to safety.

L

Liposome

A spherical vesicle made of lipid bilayers, similar to cell membranes, used to encapsulate drugs. Liposomes can protect drugs from degradation, improve targeting to specific cells (such as macrophages in tuberculosis), and enable controlled release.

M

Multimodal Antimicrobial Action

A mechanism in which a treatment attacks microbes through multiple pathways simultaneously–such as membrane disruption, oxidative stress, protein binding, and DNA interference–making resistance harder to develop.

N

Nanomaterial

A nanomaterial is a material engineered at the nanometer scale, where particles exhibit colloidal properties (typical behavior exhibited by objects with sizes ranging from 1 to 1000 nm), which explains why unique physical, chemical, or biological properties emerge at nanoscale.

Nanomedicine

Nanomedicine applies nanotechnology to healthcare, including nanoparticle-based drugs, diagnostics, vaccines, and medical devices designed for improved precision and safety.

Nanoparticle Corona (Biological or Organic Corona)

A layer of biological molecules that adsorb onto the surface of nanoparticles when they interact with biological environments. This corona determines how nanoparticles behave inside the body, influencing toxicity, targeting, and immune recognition.

Nanotoxicology

The scientific field that studies how nanomaterials interact with living systems and whether those interactions can cause harm. Unlike traditional toxicology, nanotoxicology considers size, shape, surface chemistry, and persistence as critical determinants of biological response. It asks not only "Is it toxic?" but "Where does it go, how long does it stay, and what does it trigger along the way?"

O

Oxidative Stress

A state where cells experience excessive reactive oxygen species (ROS) that damage proteins, lipids, and DNA. Some nanoparticles intentionally induce oxidative stress in microbes or cancer cells, but uncontrolled oxidative stress can harm healthy tissues.

P

Persistent Nanomaterials

Nanomaterials that do not readily degrade or dissolve in biological environments. While persistence can improve therapeutic duration, it also raises concerns about long-term safety and environmental impact.

Photothermal Therapy

A treatment approach in which nanoparticles convert light (often near-infrared) into heat, selectively killing diseased cells such as tumors, while sparing surrounding tissue.

Pilot Purgatory

Pilot purgatory describes the state in which promising innovations remain stuck in small-scale pilot studies and never progress to full adoption due to regulatory, financial, or translational barriers.

Polymeric Nanoparticles

Nanoparticles made from synthetic or natural polymers. Their safety depends on polymer chemistry, degradation products, and molecular weight. Biodegradable polymers generally show favorable nanotoxicological profiles.

Precision Medicine

Precision medicine tailors treatments to individual patients or disease characteristics. Nanotechnology enables precision by targeting drugs to specific tissues or cells.

Programmable Nanomaterials

Nanomaterials designed to respond predictably to biological or environmental cues. In green synthesis, plant-derived molecules can "program" nanoparticles to behave in biologically favorable ways.

Protein Corona

A layer of blood proteins that spontaneously coats nanoparticles once they enter the body. This "biological identity" can differ from the nanoparticle's engineered surface and strongly affects how immune cells recognize, transport, or clear it. The protein corona can reduce toxicity, increase uptake, or redirect biodistribution.

R

Reactive Oxygen Species (ROS)

Highly reactive molecules derived from oxygen. In nanomedicine, ROS can be therapeutic, by damaging pathogens or tumors, or toxic, if generated in healthy tissues. Nanotoxicology studies how ROS production depends on particle composition and environment.

Regulatory Pathway

The regulatory pathway is the process through which products are evaluated and approved by authorities to ensure safety and effectiveness. Navigating this pathway requires specialized knowledge.

Regulatory Science

The discipline that ensures medical products meet safety, quality, and efficacy standards before reaching patients. Regulatory science bridges laboratory innovation and real-world clinical use.

Reproducibility Crisis

The reproducibility crisis refers to the widespread inability to reproduce scientific findings, often due to poor documentation, variability in materials, or incomplete reporting.

Risk–Benefit Balance

The ethical and clinical evaluation of whether the expected benefits of a nanotechnology outweigh its potential risks, especially when alternatives are limited or unavailable.

S

Safe-by-Design

An approach to nanotechnology where safety considerations are integrated from the earliest design stages, rather than assessed only after development. This includes material choice, size control, degradation planning, and exposure modeling.

Scalability

Scalability is the ability to increase production from laboratory scale to industrial scale without losing quality, safety, or performance. Many innovations fail because scalability was not considered early in research.

Selective Toxicity

The principle that a therapeutic agent should harm the target (pathogen or tumor) more than healthy cells. Nanotoxicology seeks to maximize selective toxicity while minimizing collateral damage.

Silver Ion (Ag^+)

A positively charged silver atom responsible for antimicrobial activity. Controlled release of Ag^+ from silver nanoparticles enhances antimicrobial effects while reducing toxicity.

Silver Nanoparticles (AgNPs)

Metal-based nanoparticles with antimicrobial properties. Their toxicity depends on particle size, surface chemistry, silver ion release, and organic capping agents. Green-synthesized AgNPs often show improved safety profiles.

Sustainability

Sustainability refers to developing technologies and systems that meet present needs without compromising future generations. In science, this includes environmentally responsible methods and local capacity-building.

Sustained Immunity

Long-lasting immune protection achieved through controlled antigen exposure over time. Hydrogels and nanoparticle systems can extend immune stimulation beyond a single injection.

T

Time-Dependent Toxicity

Toxic effects that emerge only after prolonged exposure or delayed accumulation. Nanotoxicology emphasizes long-term observation because some adverse effects are not immediately apparent.

Translational Bottleneck

A translational bottleneck is a point where innovation stalls between discovery and application. Common bottlenecks include regulatory complexity, manufacturing challenges, and lack of translational expertise.

Translational Development

Translational development focuses on systematically moving discoveries toward real-world use by integrating science, regulation, manufacturing, clinical testing, and market considerations.

Translational Justice

The principle that scientific advances should be safe, accessible, and beneficial across populations, not only in high-resource settings. Nanotoxicology plays a critical role in ensuring that invisible fixes do not create invisible harms.

Translational Safety

The process of ensuring that safety observed in laboratory studies remains valid when scaled, manufactured, regulated, and deployed in real populations. Translational safety is essential for equitable global implementation.

V

Viral Nanoparticle (VNP)

A virus-derived structure engineered to deliver drugs, stimulate immune responses, or serve as imaging agents. VNPs are noninfectious to mammals but retain precise biological interactions. Nanostructures derived from viruses but lacking infectious genetic material. They can be highly immunogenic and useful for vaccines or cancer immunotherapy. Nanotoxicology focuses on immune overstimulation and biodistribution.

W

Waste Valorization

Waste valorization converts biological or industrial waste into valuable products. In green nanotechnology, agricultural waste can become functional nanomaterials.

Waste-to-Value Nanotechnology

An approach that converts agricultural or biological waste into valuable nanomaterials, promoting sustainability, local production, and environmental stewardship.

Z

Zoonotic Disease

A zoonotic disease is transmitted from animals to humans. Addressing such diseases requires integrated approaches like One Health.

CLOSING STATEMENT

This glossary reflects the philosophy of this book: that understanding is power, and that making science accessible is itself an act of equity. The glossary is intended to help readers from all backgrounds navigate the scientific, translational, and policy language used throughout the book. Terms are explained in plain language, with context and examples, to support understanding rather than technical mastery.

KEY READINGS

I. Foundational & Visionary Books

These works provide the intellectual and ethical backbone for understanding science, innovation, equity, and global systems.

Christensen, C.M. (2013). *The Innovator's Dilemma: When New Technologies Cause Great Firms to Fail.* Boston, MA: Harvard Business School Press. ISBN: 9781422197585.

Collins, F.S. (2009). *The Language of Life: DNA and the Revolution in Personalized Medicine.* New York, NY: HarperCollins. ISBN: 9780061963490.

Contera, S. A. (2019). *Nano Comes to Life: How Nanotechnology Is Transforming Medicine and the Future of Biology.* Princeton, NJ: Princeton University Press. ISBN: 9780691168807.

Farmer, P. (2003). *Pathologies of Power: Health, Human Rights, and the New War on the Poor* (2nd ed.). Berkeley, CA: University of

California Press. ISBN: 9780520929340

Feynman, R.P. (1992). *There's Plenty of Room at the Bottom: An Invitation to Enter a New Field of Physics* (Reprint of key published lecture in *Engineering and Science*). Pasadena, CA: California Institute of Technology; Los Angeles, CA: University of California–note: originally published as a lecture in *Engineering and Science* journal. Lecture format; archival citation: *Engineering and Science*, vol. 23, no. 5 (February 1960), pp. 22–36.

Forbath, P. (1977). *The River Congo: The Discovery, Exploration and Exploitation of the World's Most Dramatic Rivers*. New York, NY: Harper & Row. ISBN 9780061224904.

Gawande, A. (2014). *Being Mortal: Medicine and What Matters in the End*. New York, NY: Metropolitan Books. Henry Holt and Company. ISBN: 9780805095159

Kuhn, T.S. (2012). *The Structure of Scientific Revolutions*. Chicago, IL: University of Chicago Press. ISBN: 9780226458144

Marmot, M. (2015). *The Health Gap: The Challenge of an Unequal World*. London, UK: Bloomsbury Publishing. ISBN: 9781408857984

Mazzucato, M. (2018). *The Value of Everything: Making and Taking in the Global Economy*. London, UK: PublicAffairs. ISBN: 9781610396752

Mazzucato, M. (2021). *Mission Economy: A Moonshot Guide to Changing Capitalism*. London, UK: HarperCollins.

ISBN: 9780063046269

McLuhan, M. (1962). *The Gutenberg Galaxy: The Making of Typographic Man*. Toronto, ON, Canada: University of Toronto Press. ISBN: 9780802060419

Pisano, G.P. (2019). *Creative Construction: The DNA of Sustained Innovation*. New York, NY: PublicAffairs. ISBN: 9781610398763

Sachs, J.D. (2015). *The Age of Sustainable Development*. New York, NY: Columbia University Press. ISBN: 9780231173155.

Sen, A. (1999). *Development as Freedom*. New York, NY: Knopf / Oxford: Oxford University Press. ISBN: 9780198297581

Taleb, N.N. (2012). *Antifragile: Things That Gain from Disorder*. New York, NY: Random House. ISBN: 9780679645276

Yunus, M. (Contributor Weber, K.) (2007). *Creating a World Without Poverty: Social Business and the Future of Capitalism*. New York, NY: PublicAffairs. ISBN: 9781586486679

II. Nanotechnology, Nanomedicine & Delivery Systems (Reviews)

High-impact reviews that contextualize **liposomes, viral nanoparticles, hydrogels, and polymeric systems.**

Allen, T.M. & Cullis, P.R. (2013). Liposomal drug delivery systems:

from concept to clinical applications. *Advanced Drug Delivery Reviews*, 65, pp. 36–48. DOI: 10.1016/j.addr.2012.09.037

Blanco, E., Shen, H. & Ferrari, M. (2015). Principles of nanoparticle design for overcoming biological barriers to drug delivery. *Nature Biotechnology*, 33(9), pp. 941–951. DOI: 10.1038/nbt.3330

del Barrio, J., Appel, E.A., Loh, X.J. & Scherman, O.A. (2012). Supramolecular hydrogels. In *Polymeric and Self-Assembled Hydrogels*, RSC Publishing, Ed. 2012 ebook collection , 2011-2015 materials and nanoscience subject collection. https://doi.org/10.1039/9781849735629-00039.

Irvine, D.J. & Dane, E.L. (2020). Enhancing cancer immunotherapy with nanomedicine. *Nature Reviews Immunology*, 20, pp. 321–334. https://doi.org/10.1038/s41577-019-0269-6

Mitragotri, S., Burke, P. & Langer, R. (2014) Overcoming the challenges in administering biopharmaceuticals: formulation and delivery strategies. *Nature Reviews Drug Discovery* 13, 655–672. https://doi.org/10.1038/nrd4363

Nkanga, C.I., Bapolisi, A.M., Okafor, N.I., Krause, R.W.M. (2019) "General perception of liposomes: formation, manufacturing and applications". In *Liposomes - Advances and Perspectives*, ISBN 978-1-78984-495-5. Editor Angel Catala, IntechOpen, London, England. DOI: http://dx.doi.org/10.5772/intechopen.84255

Nkanga, C.I. and Nicole F. Steinmetz (2021) The pharmacology of plant virus nanoparticles. *Virology* 556, 39–61.

https://doi.org/10.1016/j.virol.2021.01.012

Nkanga, C.I., Fisch, A., Rad-Malekshahi, M., Romic, M.D., Ullrich, T., Wang, J., Kittel, B., Krause, RWM, Adler, S., Lammers, T., Hennink, W.E., Ramazani, F. (2020) Clinically established biodegradable long acting injectables: An industry perspective. *Advanced Drug Delivery Reviews* 167, 19-46. doi: 10.1016/j.addr.2020.11.008.

Pallares, R. M., Barmin, R. A., Wang, A., Kiessling, F., & Lammers, T. (2025). Clinical cancer nanomedicines. *Journal of ControlledRelease*,385,113991.Doi:10.1016/j.jconrel.2025.113991

Peer, D., Karp, J. M., Hong, S., Farokhzad, O. C., Margalit, R., & Langer, R. (2020). Nanocarriers as an emerging platform for cancer therapy. In *Nano-enabled medical applications*, Ed. Jenny Stanford Publishing. Pp 61-91.

Peer, D., Karp, J., Hong, S. et al. (2007) Nanocarriers as an emerging platform for cancer therapy. *Nature Nanotech* 2, 751–760. DOI: 10.1038/nnano.2007.387

Safari, J.B., Maseko, P., Lokole, P.B., Byamungu, G.G., Nogqala, S., Ndongwe, T., Mukanya, J.M., Ssengooba, F., Matlou, G.G., Siwe-Noundou, X., Nkanga, C.I., Pletschke, B.I., Krause, R.W.M. (2024) The Art of Liposome Surface Decoration for Targeted Drug Delivery. In *Liposomes - A Modern Approach in Research*. Editor Dr. Benjamin S. Weeks, IntechOpen, London, England. DOI: 10.5772/intechopen.1006323.

Steinmetz, N. F. (2010). Viral nanoparticles as platforms for next-generation therapeutics and imaging devices. *Nanomedicine: Nanotechnology, Biology and Medicine*, 6(5), 634-641. https://doi.org/10.1016/j.nano.2010.04.005

Steinmetz, N.F. & Evans, D.G. (2007). Utilization of plant viruses in bionanotechnology. *Organic & Biomolecular Chemistry*, 5(18), pp. 2891–2902. DOI: 10.1039/b708175h

Torchilin, V.P. (2005). Recent advances with liposomes as pharmaceutical carriers. *Nature Reviews Drug Discovery*, 4(2), pp. 145–160. DOI: 10.1038/nrd1632

Torchilin, Vladimir P. (2014) "Multifunctional, stimuli-sensitive nanoparticulate systems for drug delivery." *Nature reviews Drug discovery* 13(11): 813-827. DOI: 10.1038/nrd4333

Lammers, T. (2024). Nanomedicine tumor targeting. *Advanced Materials*, 36(26), 2312169. https://doi.org/10.1002/adma.202312169

Lammers, T., Aime, S., Hennink, W. E., Storm, G., & Kiessling, F. (2011). Theranostic nanomedicine. *Accounts of chemical research*, 44(10), 1029-1038. https://doi.org/10.1021/ar200019c

Lammers, T., Kiessling, F., Ashford, M., Hennink, W., Crommelin, D., & Storm, G. (2016). Cancer nanomedicine: is targeting our target?. *Nature Reviews Materials*, 1(9), 16069. https://doi.org/10.1038/natrevmats.2016.69

Wang, Y., Kohane, D. (2017) External triggering and triggered targeting strategies for drug delivery. *Nat Rev Mater* 2, 17020. https://doi.org/10.1038/natrevmats.2017.20

Witika, B. A., Poka, M. S., Makoni, P. A., Ramazani, F., & Nkanga, C. I. (2025). Clinical Applications of Long Acting Injectables: Systemic and Local Drug Delivery. In *Biodegradable Long Acting Injectables and Implants: An Industry View* (pp. 185-220). Cham: Springer Nature Switzerland. https://doi.org/10.1007/978-3-031-72302-5_7.

Wolfram, J., & Ferrari, M. (2019). Clinical cancer nanomedicine. *Nanotoday*,25,85-98.https://doi.org/10.1016/j.nantod.2019.02.005

III. Antimicrobial Resistance & Alternative Strategies

Core literature framing AMR as a global systems failure.

Antimicrobial Resistance Collaborators (Mohsen Naghavi et al.) (2016). Global burden of bacterial antimicrobial resistance in 2019: a systematic analysis. *The Lancet* 399(10325): 629–655. doi: 10.1016/S0140-6736(21)02724-0

Davies, J. & Davies, D. (2010). Origins and evolution of antibiotic resistance. *Microbiology and Molecular Biology Reviews*, 74(3), pp. 417–433. https://doi.org/10.1128/mmbr.00016-10

Huh, A. & Kwon, Y. (2011). "Nanoantibiotics": A new paradigm for treating infectious diseases using nanomaterials in the antibiotic resistant era. *Journal of Controlled Release*, 156(1), pp. 128–145. https://doi.org/10.1016/j.jconrel.2011.07.002

Laxminarayan, R., Duse, A., Wattal, C., Zaidi, A.K.M., Wertheim, H.F.L., Sumpradit, N., et al. (2013). Antibiotic resistance–the need for global solutions. *The Lancet Infectious Diseases*, 13(12), pp. 1057–1098. https://doi.org/10.1016/S1473-3099(13)70318-9

O'Neill, J. (2014) Antimicrobial Resistance: Tackling a crisis for the health and wealth of nations. London, UK: Review on Antimicrobial Resistance. Available at https://wellcomecollection.org/works/rdpck35v/items?shouldScrollToCanvas=true

O'Neill, J. (2016). *Tackling Drug-Resistant Infections Globally: Final Report and Recommendations.* Review on Antimicrobial Resistance. UK Government. Available at: https://wellcomecollection.org/works/rdpck35v

Rai, Mahendra, Alka Yadav, and Aniket Gade (2009) Silver nanoparticles as a new generation of antimicrobials. *Biotechnology advances* 27(1): 76-83. https://doi.org/10.1016/j.biotechadv.2008.09.002

Ventola, C.L. (2015). The antibiotic resistance crisis: part 1: causes and threats. *Pharmacy and Therapeutics*, 40(4), pp. 277–283. PMID: 25859123

World Health Organization (2015). Global action plan on antimicrobial resistance. World Health Organization. Geneva.

ISBN 9789241509763. Available at: https://iris.who.int/handle/10665/193736

IV. Green Nanotechnology & Bio-Hybrid Systems

Key works underpinning plant-mediated synthesis, sustainability, and local innovation.

Anandalakshmi, K. (2021) Green synthesis of silver nanoparticles using plant extracts–a review." *Plant Archives* 21(1): 2091-2097.

Anastas, P.T. & Warner, J.C. (2000). *Green Chemistry: Theory and Practice*. Oxford: Oxford University Press.

Ghaffari-Moghaddam, M., Hadi-Dabanlou, R., Khajeh, M. et al. (2014) Green synthesis of silver nanoparticles using plant extracts. *Korean J. Chem. Eng.* 31, 548–557. https://doi.org/10.1007/s11814-014-0014-6

Iravani, S. (2011). Green synthesis of metal nanoparticles using plants. *Green Chemistry*, 13 (10), pp. 2638–2650. DOI: 10.1039/c1gc15386b

Kabeya, Jimmy K., et al. (2025) "Antimicrobial capping agents on silver nanoparticles made via green method using natural products from banana plant waste." *Artificial Cells, Nanomedicine, and Biotechnology* 53.1: 29-42. https://doi.org/10.1080/21691401.2025.2462335

Kambale, Espoir K. et al. (2020) 'Green synthesis of antimicrobial silver nanoparticles using aqueous leaf extracts from three Congolese plant species (Brillantaisia patula, Crossopteryx febrifuga and Senna siamea), *Heliyon*, 6, e04493.

Myriam M. Luzala, Claude K. Muanga, Joseph Kyana, Justin B. Safari, Eunice N. Zola, Grégoire V. Mbusa, Yannick B. Nuapia, Jean-Marie I. Liesse, Christian I Nkanga, et al.(2022) A critical insight into the evaluation of antimicrobial and anti-biofilm activities of green synthesized plant-based metallic nanoparticles. *Nanomaterials* 12(11), 1841; https://doi.org/10.3390/nano12111841.

V. Vaccines, Immunotherapy & Viral Nanoparticles

Bachmann, M.F. & Jennings, G.T. (2010). Vaccine delivery: a matter of size, geometry, kinetics and molecular patterns. *Nature Reviews Immunology*, 10(11), pp.787–796. https://doi.org/10.1038/nri2868

Chung, Y. H., Cai, H., & Steinmetz, N. F. (2020). Viral nanoparticles for drug delivery, imaging, immunotherapy, and theranostic applications. *Advanced drug delivery reviews,* 156, 214-235. https://doi.org/10.1016/j.addr.2020.06.024

Irvine, D.J., Hanson, M.C. & Rakhra, K. (2015) Synthetic nanoparticles for vaccines and immunotherapy." *Chemical reviews* 115.19: 11109-11146. https://doi.org/10.1021/acs.chemrev.5b00109

Nkanga, C.I., and Steinmetz, N.F. (2022) Injectable hydrogel containing cowpea mosaic virus nanoparticles prevents colon cancer growth. *ACS biomaterials science & engineering* 8.6 (2022): 2518-2525. https://doi.org/10.1021/acsbiomaterials.2c00284

Nkanga, C.I., Ortega-Rivera, O.A., Shin, M.D., Moreno-Gonzalez, M., and Steinmetz, N.F. (2022) Injectable slow-release hydrogel formulation of a plant virus-based COVID-19 vaccine candidate, *Biomacromolecules*, 23(4), pp. 1812–1825. https://doi.org/10.1021/acs.biomac.2c00112

Pardi, N., Hogan, M.J., Porter, F.W. & Weissman, D. (2018). mRNA vaccines–a new era in vaccinology. *Nature Reviews Drug Discovery*,17(4),pp.261–279.https://doi.org/10.1038/nrd.2017.243

VI. Nanotoxicology & Safety

Essential for responsible innovation and translational trust.

Nel, A., Xia, T., Mädler, L. & Li, N. (2006). Toxic potential of materials at the nanolevel. *Science*, 311(5761), pp. 622–627. https://doi.org/10.1126/science.1114397

Nkanga, C.I. (2023) Nanotoxicological profiles of clinically approved nanoplatforms. *Beni-Suef Univ. J. Basic. Appl. Sci.* 12, 115. https://doi.org/10.1186/s43088-023-00458-6

Nkanga, C.I. and Nicole F. Steinmetz (2021) The pharmacology of plant virus nanoparticles. *Virology* 556, 39–61. https://doi.org/10.1016/j.virol.2021.01.012

Nyström, Andreas M., and Bengt Fadeel (2012) Safety assessment of nanomaterials: implications for nanomedicine. *Journal of controlled release* 161(2)403-408. https://doi.org/10.1016/j.jconrel.2012.01.027

Oberdörster, G., Oberdörster, E. & Oberdörster, J. (2005). Nanotoxicology: an emerging discipline evolving from studies of ultrafine particles. *Environmental Health Perspectives*, 113(7), pp. 823–839. https://doi.org/10.1289/ehp.7339

Witika, B.A.; Makoni, P.A.; Matafwali, S.K.; Chabalenge, B.; Mwila, C.; Kalungia, A.C.; Nkanga, C.I.; et al. (2020) Biocompatibility of Biomaterials for Nanoencapsulation: Current Approaches. *Nanomaterials* 10, 1649. https://doi.org/10.3390/nano10091649

VII. Translational Science, CDMOs & Regulation

Balogun, T. et al. (2022). *Why health-tech startups fail.* Industry Reports. https://healthjournal.uk/health-techs-hidden-pitfalls-why-brilliant-ideas-still-flop/

Collins, Francis S. (2011) Reengineering translational science: the time is right. *Science translational medicine* 3(90): 90cm17. DOI: 10.1126/scitranslmed.3002747

European Medicines Agency (EMA) (2006). *Reflection Paper on the Regulatory Requirements for Nanomedicines*. Pre-authorisation Evaluation of Medicines for Human Use. EMA, Amsterdam. London, 29 June 2006 EMEA/CHMP/79769/2006

Food and Drug Administration (FDA) (2014). *Guidance for Industry: Considering Whether an FDA-Regulated Product Involves the Application of Nanotechnology.* U.S. FDA Docket No. FDA-2010-D-0530. https://www.federalregister.gov/documents/2014/06/27/2014-15033/guidance-for-industry-considering-whether-a-food-and-drug-administration-regulated-product-involves

Halwani, Abdulrahman A. (2022) Development of pharmaceutical nanomedicines: from the bench to the market." *Pharmaceutics* 14(1): 106. https://doi.org/10.3390/pharmaceutics14010106

Mangla, Bharti, et al. (2025) Regulating nanomedicines: challenges, opportunities, and the path forward." *Nanomedicine* 20(15): 1911-1927. https://doi.org/10.1080/17435889.2025.2533107

Pallares, R. M., Barmin, R. A., Wang, A., Kiessling, F., & Lammers, T. (2025). Clinical cancer nanomedicines. *Journal of Controlled Release,* 385, 113991. Doi: 10.1016/j.jconrel.2025.113991

Pisano, G.P. (2006). *Science Business: The Promise, the Reality, and the Future of Biotech*. Boston, MA: Harvard Business School Press. ISBN: 9781591398400

STATNANO. Nanotechnology Publications: Number and Annual Growth Rate of Nano-articles (2000-2022) | https://share.google/W7KIXR5uMtgoBHrnT

Thapa, Raj Kumar, and Jong Oh Kim (2023) Nanomedicine-based commercial formulations: current developments and future prospects. *Journal of Pharmaceutical Investigation* 53.1: 19-33. doi:

10.1007/s40005-022-00607-6

VIII. Brain Circulation, Equity & Global Science

African Academy of Sciences (2024). *Research Capacity in Africa.* AASReports.Availableat:https://www.aasciences.africa/?q=content/research-capacity-strengthening-reports

Agrawal, Ajay, et al. (2011) Brain drain or brain bank? The impact of skilled emigration on poor-country innovation." *Journal of Urban Economics* 69(1): 43-55. https://doi.org/10.1016/j.jue.2010.06.003

Agrawal, Ajay. (2014) *Diaspora networks, knowledge flows and brain drain*. Vol. 15. World Intellectual Property Organization, Ajay Agrawal (WIPO), 2014.

Batista, Catia, et al. (2025) Brain drain or brain gain? Effects of high-skilled international emigration on origin countries. *Science* 388 (6749): eadr8861. DOI: 10.1126/science.adr8861

Bhardwaj, Bhawana, and Dipanker Sharma (2023) Migration of skilled professionals across the border: Brain drain or brain gain?." *European Management Journal* 41.6: 1021-1033. https://doi.org/10.1016/j.emj.2022.12.011

Collins, Francis S. (2011) Reengineering translational science: the time is right. *Science translational medicine* 3(90): 90cm17.

Ito, T., Kondo, R. & Ueda, S. (2025). Brain bridges: International research collaboration and development outcomes. *Journal of Development Economics*, 177, 102728.

Ito, R, D Chavarro, T Ciarli, R Cowan, and F Visentin (2025) From drains to bridges: The role of internationally mobile PhD students in linking non-mobile with foreign scientists," *Journal of Development Economics* 177: 103577. https://doi.org/10.1016/j.jdeveco.2025.103577

Saxenian, A. (2005). From brain drain to brain circulation: Transnational communities and regional upgrading in India and China. *Studies in Comparative International Development*, 40, pp. 35–61. https://doi.org/10.1007/BF02686293

IX. Author's Selected Original Contributions

Daramola, O.A., Siwe-Noundou, X.S., Nkanga, C.I., Tseki, P.F. and Krause, R.W.M. (2020) 'Synthesis of pH-sensitive dual capped CdTe quantum dots: their optical properties and structural morphology', *Journal of Fluorescence*, 30, pp. 557–564.

Kabeya, J.K., Ngombe, N.K., Mutwale, P.K., Safari, J.B., Matlou, G.G., Krause, R.W.M. and Nkanga, C.I. (2025) 'Antimicrobial capping agents on silver nanoparticles made via green method using natural products from banana plant waste', *Artificial Cells, Nanomedicine, and Biotechnology*, 53(1), pp. 29–42.

Kabayo, S.M., Kindala, J.T., Nkanga, C.I., Krause, R.W.M. and Taba, K.M. (2019) 'Preparation and characterization of solid acid catalysts derived from coffee husks', *International Journal of Chemical Science*, 3(6), pp. 05–13.

Kambale, E. K., Nkanga, C. I., Mutonkole, B. P. I., Bapolisi, A. M., Tassa, D. O., Liesse, J. M. I., ... & Memvanga, P. B. (2020) 'Green synthesis of antimicrobial silver nanoparticles using aqueous leaf extracts from three Congolese plant species (Brillantaisia patula, Crossopteryx febrifuga and Senna siamea)', *Heliyon*, 6(8).

Lokole, P.B., Ngombe, N.K., Motomba, D.I., Safari, J.B., Mpuza, M.K., Krause, R.W.M., Mutwale, P.K. and Nkanga, C.I. (2024) 'Preparation and characterization of micellar nanoparticles using crude saponins from five Congolese plant species', *Pharmaceutical Science Advances*, 2, 100055.

Mukubwa, G.K., Nkanga, C.I., Buya, A.B., Mbinze, J.K., Krause, R.W.M. and Memvanga, P.B. (2020) 'Self-nanoemulsifying drug delivery systems (SNEDDS) for oral delivery of Garcinia kola seeds ethanolic extract: formulation, characterization and in vivo antimalarial activity', *Journal of Pharmacy and Pharmacognosy Research*, 8(3), pp. 177–190.

Mukaya, D. W., Makuala, F. M., Muipata, M. M., Modia, G. M., Balwanga, C., Nangulu, C., ... & Nkanga, C. I. (2026). 'Nanoencapsulation and antimicrobial activities of Commiphora swynnertonii resin against multidrug-resistant Staphylococcus aureus strains from mastitis cow patients'. *Discover Nano*, 21(1), 63.

Nkanga, C.I. and Krause, R.W.M. (2018) 'Conjugation of isoniazid to a zinc phthalocyanine via hydrazone linkage for pH-dependent liposomal controlled release', *Applied Nanoscience*, 8(6), pp. 1313–1323.

Nkanga, C.I. and Krause, R.W.M. (2019) 'Encapsulation of isoniazid-conjugated phthalocyanine-in-cyclodextrin-in-liposomes using heating method', *Scientific Reports*, 9(1), 11485.

Nkanga, C.I., and Steinmetz, N.F. (2022) Injectable hydrogel containing cowpea mosaic virus nanoparticles prevents colon cancer growth. *ACS biomaterials science & engineering* 8.6 (2022): 2518-2525.

Nkanga, C.I., Ortega-Rivera, O.A., Shin, M.D., Moreno-Gonzalez, M., and Steinmetz, N.F. (2022) 'Injectable slow-release hydrogel formulation of a plant virus-based COVID-19 vaccine candidate', *Biomacromolecules*, 23(4), pp. 1812–1825.

Nkanga, C.I., Ortega-Rivera, O.A. and Steinmetz, N.F. (2022) 'Photothermal immunotherapy of melanoma using TLR-7 agonist laden tobacco mosaic virus with polydopamine coat', *Nano-medicine: Nanotechnology, Biology and Medicine*, 44, 102573. https://doi.org/10.1016/j.nano.2022.102573

Nkanga, C.I., Roth, M., Walker, R.B. and Krause, R.W.M. (2020) 'Co-loading of isoniazid-grafted phthalocyanine-in-cyclodextrin and rifampicin in crude soybean lecithin liposomes: formulation, spectroscopic and biological characterization', *Journal of Biomedical Nanotechnology*, 16(1), pp. 14–28.

Nkanga, C.I., Shukla, S., Zhou, J., Chung, Y.H., Jokerst, J.V. and

Steinmetz, N.F. (2021) 'The in vivo fate of tobacco mosaic virus nanoparticles theranostic agents modified by the addition of a polydopamine coat', *Biomaterials Science,* 9, pp. 7134–7150.

Nkanga, C.I., Walker, R.B., Noundou, X.S. and Krause, R.W.M. (2019) 'Co-encapsulation of rifampicin and isoniazid in crude soybean lecithin liposomes', *South African Journal of Chemistry,* 72(1), pp. 80–87.

Nkanga, C.I., Walker, R.B. and Krause, R.W.M. (2018) 'pH-dependent release of isoniazid from isonicotinic acid (4-hydroxy-benzylidene)-hydrazide loaded liposomes', *Journal of Drug Delivery Science and Technology*, 45, pp. 264–271.

Nkanga, C.I., Krause, R.W.M., Noundou, X.S. and Walker, R.B. (2017) 'Preparation and characterization of isoniazid loaded liposomes made of crude soybean lecithin', I*nternational Journal of Pharmaceutics*, 526(1), pp. 466–473.

Nwahara, N., Niaz, S., Managa, M., Nkanga, C. I., Adegoke, O., & Achadu, O. J. (2026). 'Nanoplasmonic core–shell nanoraspberry chip for ultrasensitive surface-enhanced Raman scattering detection of SARS-CoV-2: a modular nanobiosensor for respiratory virus diagnostics'. *Sensors & Diagnostics*. DOI: 10.1039/d5sd00185d

Okafor, N.I., Nkanga, C.I., Noundou, X.S., Walker, R.B. and Krause, R.W.M. (2019) 'Encapsulation and physicochemical evaluation of efavirenz in liposomes', *Journal of Pharmaceutical Investigation*, 50, pp. 201–208.

Safari, J.B., Mona, L.B., Sekaleli, B.T., Avudi, B.K., Isamura, B.K., Mukubwa, G.K., Aderotimi, S.S., Mbinze, J.K., Lobb, K.A., Krause, R.W.M. and Nkanga, C.I. (2023) 'Inclusion complexation and liposomal encapsulation of an isoniazid hydrazone derivative in cyclodextrin for pH-dependent controlled release', *Journal of Drug Delivery Science and Technology*, 81, 104302.

Wu, Z., Zhou, J., Nkanga, C.I., Jin, Z., He, T., Borum, R.M., Yim, W., Zhou, J., Cheng, Y., Xu, M., Steinmetz, N.F. and Jokerst, J.V. (2022) 'One-step supramolecular multifunctional coating on plant virus nanoparticles for bioimaging and therapeutic applications', *ACS Applied Materials & Interfaces*, 14(11), pp. 13692–13702.

www.ingramcontent.com/pod-product-compliance
Lightning Source LLC
LaVergne TN
LVHW040517210926
848480LV00007BA/1494